AIRFIELDS
AND LANDING GROUNDS OF WALES: NORTH

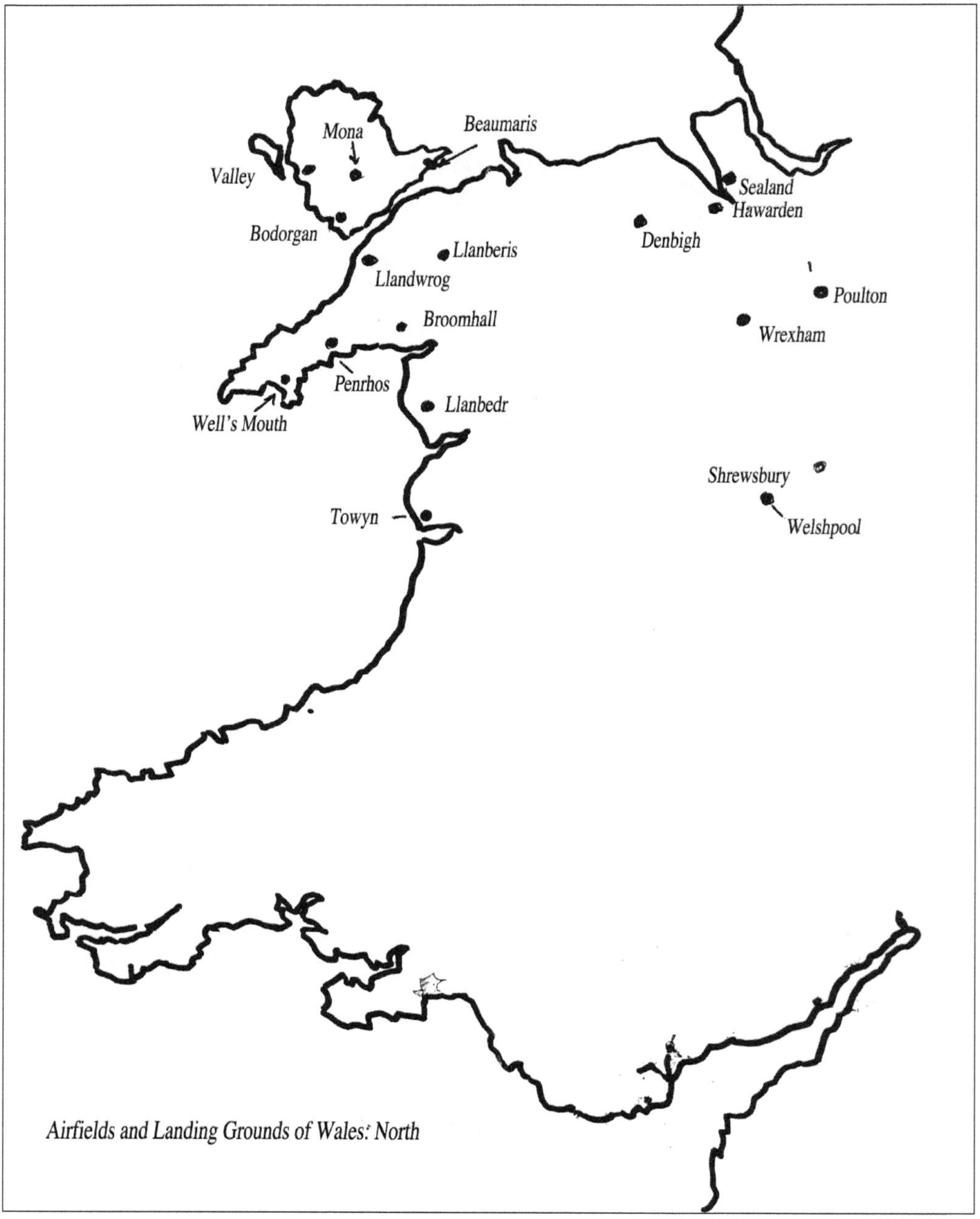

Airfields and Landing Grounds of Wales: North

AIRFIELDS

AND LANDING GROUNDS OF WALES: NORTH

IVOR JONES

The History Press

First published 2008

The History Press
The Mill, Brimscombe Port
Stroud, Gloucestershire, GL5 2QG
www.thehistorypress.co.uk

British Library Cataloguing in Publication Data.
A catalogue record for this book is available from the British Library.

ISBN 978 0 7524 4510 6

Typesetting and origination by The History Press Ltd

Contents

Preface

As in other regions of Wales most of the airfields of North Wales are on the coast, perched on a narrow strip of flat land between the sea and mountains. The difference between this region and those further south is the size and height of the mountains. The weather is wetter and windier as one reaches Anglesey and Snowdonia, and flying was always going to be tricky for the many schools of training in this area. Llanbedr, Valley, Penrhos and Wrexham were used as fighter stations early in the Second World War, but they were to revert to training aircrews, as did the other airfields in North Wales.

Target-towing was carried out at Morfa Towyn, Llanbedr, Penrhos, Hell's Mouth, Llandwrog and Bodorgan for the many Anti-Aircraft Schools around the coast of Wales and north-west England. Flying Training took place at Valley, Sealand, Wrexham, Hawarden, Poulton and Mona.

Broomhall was, before and after the war, a private civil aerodrome. Welshpool is a recently conceived and worthy mid-Wales Airport. Post-war civil flying also takes place at Mona and Llandwrog as Caernarfon Airport.

Beaumaris and Messrs Saunders-Roe made a good team during the most depressing days of the Second World War. A need appeared and it was filled up in the Menai Straits. A quickly conceived and rapidly put to work flying boat facility was to see every Catalina brought up to Coastal Commands specifications with mostly unskilled labour. Hawarden, now re-named Broughton, is still making wings for the Airbus variants and servicing the Hawker 125.

The saddest sight of all to see in my visits to the North Wales airfields is that of RAF Wrexham. It has been quarried out of existence and a great scar replaces the old Borras airfield.

As in the previous two volumes of this series, focusing on the regions of the west and south, there were, during the last two years of the Second World War, a few US Army cubstrips in North Wales. Denbigh and Wrexham had four, along with Aberaeron at the anti-aircraft school, and they were all busy training, mostly in Air Observation Post duties, with the Field Artillery. In the south the artillery was virtually queuing up to get on to the Sennybridge gun ranges on the Brecon Beacons to get some time in at air-to-ground wireless communication and pass on the fall of shot corrections. The American artillery in the north was also waiting, I am pretty sure, for their chance on the Trawsfynedd Army Artillery Range. This was a large area of land range that lay astride the A470 south of Trawsfynydd, that extended east toward Bala and west to Penrhyndeudraeth.

I am aware of the possibility that I have missed out some airfields, especially the several light aircraft and glider fields that abound in the north-east corner of Wales. I do apologise but it is too far away to study in any depth from my base in the south.

Acknowledgements

I would like to thank the following people for their assistance in the compilation of these histories: David J. Smith, Aldon Ferguson, Medwyn Parry, Roy Sloan, E. Doylerush, Peter Waters and Roger Davies of DERA, D. Pratt and M. Grant, Ioan Gwyllim Jones of Flintshire Record Office, Steve Willis and Barry Hollis, Derek Jones, Chris Carrow, Wing Commander David Annand, Don Clarke, Raymond L. Wheeler and Eileen Simpson of the Grosvenor Estate and most of all the wonderful work in researching the RAF overhead photographs for these books carried out by Derek Elliot and his team at the Welsh Assembly Government Photograph Archive.

If any readers of these books require more information on the history of the UK's airfields then membership of the Airfield Research Group, with its magazine *ARO Review*, should be the next step.

Introduction

North Wales has seen its share of aviation progress, from its infancy to the Second World War, when it became important to the day and night defence of north-west England, the training of aircrews and the entry point from the USA for most of the heavy bombers of the US Air Force. It still holds a key position today, with RAF Valley as the main Tactical Fighter Training Unit, RAF Sealand still providing vital aeronautical equipment and, of course, the Airbus factory at Broughton producing the wings for every type and model of the Airbus series. The superb Mountain Rescue Service of the Royal Air Force that we still have today originated in the north.

The fine coast of Anglesey at Beaumaris was utilised by Saunders Roe to modernise and repair the Catalina flying boats during the war. After the war they continued production of other designs of flying boats and seaplanes for civilian use, but the day of the seaplanes was ending and the plant was to close.

As early as 1912 the pioneer of flight Gustav Hamel flew his aircraft at Wrexham Racecourse, thrilling the crowds with his displays of aerobatics, and nearby Borras Lodge field was used by Sir Alan Cobham's Flying Circus in 1932 and 1933. Borras was to develop into the wartime airfield that was RAF Wrexham.

In 1916 T. Murray Dutton built his own aircraft on his own airfield at the edge of the river Dee at Queensferry (Sealand). The objective was to form a flying school and equip it with his planes. He was so successful that the Royal Flying Corps was to take it over, lock stock and barrel, in 1917.

We can also go back to the First World War at Llangefni Anglesey, which was an airship base for the Royal Naval Air Service. The airships' job was to patrol the waters off Merseyside and the Irish Sea, looking for the U-boats of the enemy. In the Second World War the site was developed into the aircrew training station of RAF Mona and is still used today as a satellite of RAF Valley and some private club flying.

This RAF overhead shows the airfield of Towyn on 1 May 1946, when the station had closed. Note the north-north-west–south-south-east wear track due to the prevailing wind and its take-off and landing consequences. (Welsh Assembly Government Photographic Archive)

ONE
RAF Towyn (Tywyn)

Grid Reference SH 575015, Gwynedd

This small grass airfield near the sea was built for the purpose of flying targets for the anti-aircraft gunnery practice needed by the Royal Artillery Anti-Aircraft School at nearby Tonfanau, where gun teams were taught the skills of high-, medium-, and low-altitude defence against attack from the air. The batteries were equipped with 4.5in, 3.7in and Bofors 40mm low-altitude guns.

This was a premier school, with Manorbier being its only equal in the training of anti-aircraft gunners. Its guns fired out to sea, into Cardigan Bay and the Irish Sea, and the camp was large, with a capacity of up to 1,000 soldiers. It was decided that the school would need a dedicated target-towing station to handle Tonfanau.

Tonfanau housed, at one time or another, No.55 Heavy AA Regiment, No.37 Light AA Regiment and a steady influx of American AA Gunners, as well as the normal school personnel.

Morfa Towyn farmlands and farmhouse were requisitioned early in 1940 and over the whole of that year, and it took until well into 1941 to finish the job of erecting hangars, workshops, stores and living quarters. There were a lot of other military training units in the area. The Royal Marines had occupied several large houses in Towyn and No.22 Royal Marine Commando training camp was here. A Royal Navy hospital, the Royal Army Service Corps, Amphibious Vehicle Training Wing and a battalion of the Welch Regiment were all vying for building work to be done and sparse water, sewage, and electricity services to be spread around. The incoming airmen that were to service RAF Towyn had to go into army barracks to shelter or be billeted in farms.

The first of the aircraft arrived: Hawker Henleys flown over from RAF Penrhos across the bay. These were to be the target tugs at Towyn for a long time and were to cause some controversy as to their fitness for the job of sustained towing at fairly high speeds.

Then from St Athan came the 'Queen Bees' of the PAU (Pilotless Aircraft Unit). These were basically Tiger Moths fitted with what was very high-tech at that time: radio remote control, enabling these aircraft to be sent into the air, be directed to the target area, fly a requested course and return to a safe landing. This control could be passed over from the operator on the ground to another operator at the guns at Tonfanau a couple of miles north up the coast. It worked most of the time, but conditions had to be perfect. That was rare but fairly regular and the gunners would attempt to blast the Queen Bee from the sky.

The station at Towyn was part of No.1 AACU with its HQ at Farnborough, who came under the HQ of No.70 Group, also at Farnborough, who in turn came under Army Co-operation Command, Bracknell, Berkshire.

Later, other aircraft were brought in to do specific jobs of Army co-operation, such as Lysanders, Martinets and Hurricanes, but the Henleys went on even after other Army co-operation bases had ceased to use them. They were not designed for the work and a makeshift that suffered from engine stressing when drogue towing. There were many crashes with these aircraft at Towyn. In the opinion of those in the know the large casualty rate at Towyn was down to the Henleys. This aircraft was

The bath-house and emergency decontamination block at RAF Morfa Towyn. (I. Jones)

Frames 0180 and 0181 of a PR Sortie of 1 August 1950 when the airfield was in use as a summer camp. There is still a hint of the old grass runways showing. (Welsh Assembly Government Photographic Archive)

This Hawker Henley is shown flying over the seafront of Towyn. The same pilot and aircraft crashed the next day. (D. Holmes, via Roger Thomas)

The Queen Bee, a pilot-less derivative of the Tiger Moth, was used as a target aircraft at Towyn with some success.

designed as a light, fast bomber, with a similar engine and similar shape as the Hurricane but had a crew of two and carried towing gear and a winch which added weight. Air-to-air gunnery schools had soon worn out the engines of their Henleys and they were issued to anti-aircraft co-operation units instead. Another factor was that when towing at a speed above 200mph the engine temperature soared. The wartime pilots who were not considered up to operational flying were posted to anti-aircraft co-operation units for drogue towing duties. A further factor was that the maintenance in the hangars of operational training units was far better and the best technicians were to be found there.

The target tugs comprised 'C' Flight and the Queen Bees of 'U' Flight of No.1 AACU. (Anti-Aircraft Co-operation Unit).

A Henley crashed on the marshes, west of Escuan Farm, the engine cutting out just after take-off and slewed into a drainage ditch. The pilot was unhurt and the plane salvaged. As the winter arrived the problems that were to worry the station right through showed themselves: wet weather and mud. There was little in the way of paved taxi-ways, hard-stands and aprons or roads. This, with the fact that there were no runways, forced them to detach aircraft to nearby Llanbedr in the wettest periods.

The HQ was finished at last on 19 October 1940. The strength of the station was only twelve officers and 226 airmen.

A Beaufort L4462 made an emergency landing here on 10 November 1940 after engine trouble on a flight from St Eval to Abbotsinch. On the following day the airfield had to be declared unserviceable because of heavy rain, which effectively imprisoned the Beaufort for another two days before conditions improved sufficiently for the aircraft to depart. On the same day a Magister arrived from Penrhos to be the station hack.

The transmitter arrived that was needed to control the Queen Bees in November, but it took three months to sort out the snags before the first radio-controlled flight took place with a safety pilot on board, on 25 February 1941. The Tonfannau Range was notified that the artillery now had a target and firing at an unmanned target could commence.

The Queen Bees were successful but sophisticated and so suffered from damage to their control systems from any near miss from the guns. Losses occurred frequently, averaging one to two a month.

Henley L3284 force-landed half a mile from the airfield in June 1941, killing the pilot and the drogue operator, and then L3435, which crashed offshore near Tonfanau after engine failure in

May 1942. The drogue operator was thrown clear, but the twenty-four-year-old pilot was killed. A lot of these accidents were just after take-off.

The target towing, weather permitting, continued day after day, Sundays included. More of the buildings were handed over and quickly occupied. The Air Training Corps cadets were now more able to be accommodated.

In February 1942 two Lysanders, aiding the unit from No.6 AACU, arrived for co-operation with No.225 Battery RA and Ground Gunners Practice Camp (Vickers and Lewis guns). 10 March 1942 saw the odd event of bombing practice by Hawker Henleys.

On 18 December 1942 twelve Lockheed Lightnings of the 97th Fighter Squadron USAAF arrived unexpectedly. Whilst on route from Northern Ireland to St Eval, bad weather forced a diversion to Towyn, which itself was waterlogged. One of the planes overshot into a gun pit resulting in Category E damage (written off). The others flew off two days later, after having a ball at the local dance in the town, where they greatly impressed the girls with their dancing, and caused a stir when one of the pilots drew a revolver and fired five shots into the roof. The police and MPs took a long time to restore order!

This Lysander, belonging to No.6 Anti-Aircraft Co-operation Unit, is on the ramp outside the hangars at Towyn, on detachment there in May 1942.

Lockheed Lightnings of the 97th Fighter Squadron US Air Force at Towyn.

On 5 May 1942 a twenty-four-year-old New Zealand pilot, Sgt J.H. Wilson, experienced engine failure in his Henley. He crashed on the rocky shore north of Tonfanau and was killed.

On 1 November 1942, 'C' Flight was redesignated No.1605 AACU Flight. 'U' Flight had disbanded on 30 October 1942 at Towyn. On 23 November No.1605 Flight took delivery of three Martinets and at the same time a Henley was lost when L3324 was flown into cloud obscuring high ground near Pen-y-Groes, Caernavon. The pilot was killed instantly. On 1 February 1943 Flight Lieutenant Probyn started his spell of misfortune when landing his Hawker Henley on Towyn. Finding his brakes were ineffective owing to the surface water he was unable to prevent a collision with other aircraft parked on the airfield, writing his plane off and causing damage to others. Flt Lt Probyn was unhurt but two days later, on the 3rd, he was involved in another accident. After take-off, and climbing to 300ft, his Henley's engine cut out. He force-landed in the sea off Tonfanau Camp and was rescued. Although his aircraft was not badly damaged it was impossible to pull it to shore because of the rocky nature of the shore. The Army was called upon to blow it up, leaving the engine and propeller in the sea.

P/O George Probyn and his drogue man, Cpl Leslie Pitt, with their Henley at Towyn on 2 April 1942.

No.631 Squadron at Towyn posing on and in front of one of their Hurricanes. On the table is displayed the various types of ammunition fired by the artillery at their aircraft-towed targets.

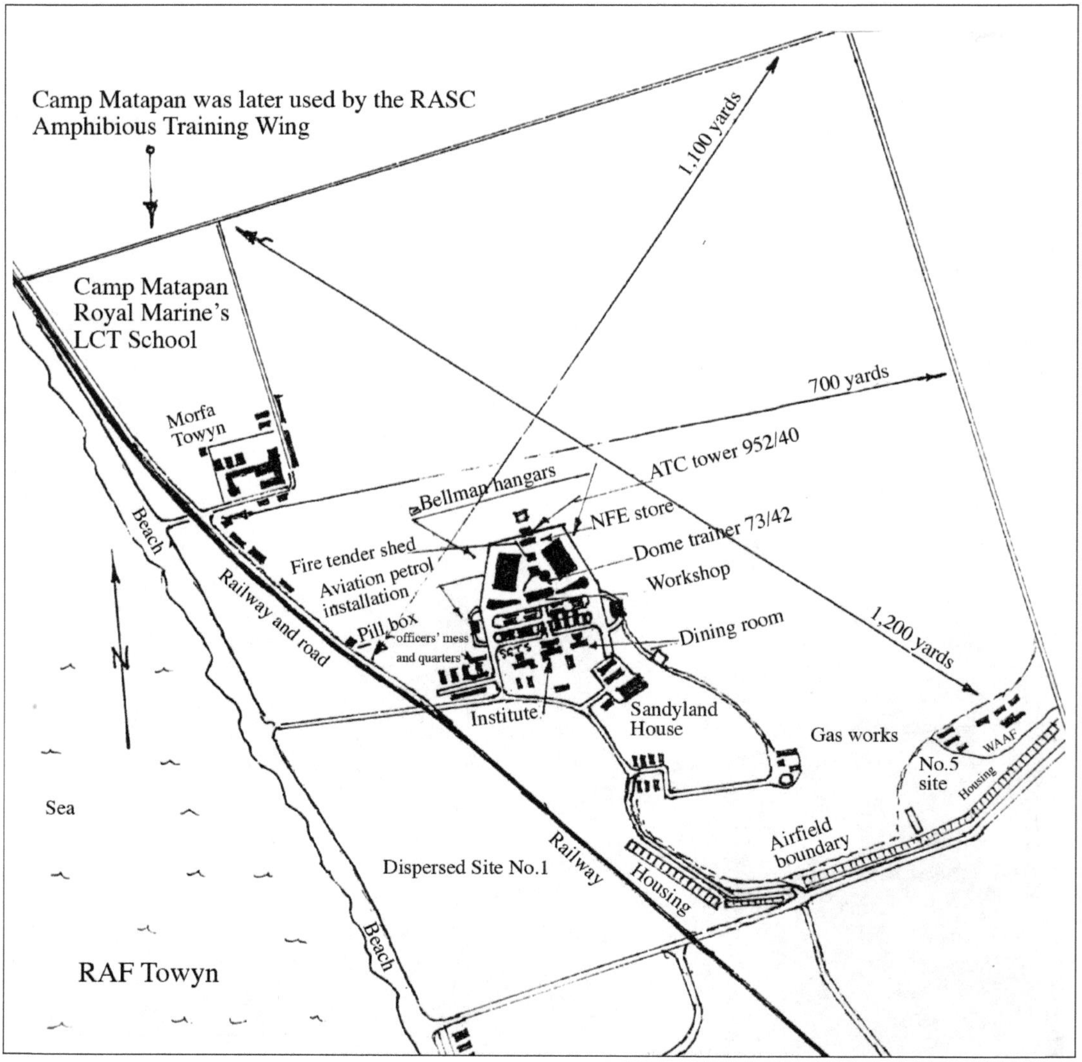

Map showing the buildings and runways at Towyn. (I. Jones)

In 1943 Army Co-operation Command was disbanded and its assets absorbed into Fighter Command, which was now responsible for Towyn. There were further changes on 15 November when Fighter Command was disbanded and its assets were to be handed to the Supreme HQ of the Allied Expeditionary Force (SHAEF). The aircraft were to be part of 2nd TAF (Tactical Air Force) and the separate Air Defence of Great Britain. This brought change. On 1 December 1943 the resident No.1605 Flight combined with No.1628 Flight to become No.631 Squadron. No.1628 Flight was formed at Towyn by redesignating No.4 RAF Regiment Anti-Aircraft Practice Camp as Target Towing Flight on 17 June 1943.

This new larger unit still used the Henley for towing, but now, with other types such as the Martinet, Lysander 2 and Vengeance V. There were also aircraft for gun laying and tracking, first with the Hurricane, then Harvard, Spitfire VB and then Spitfire 16s.

On 20 February 1944 Henley L3325 caught fire in the air and was ditched in the sea a few hundred yards off shore, 2 miles south of the airfield. Its crew was picked up by launch. One of the many Army units based nearby was the Royal Army Service Corps's No.1 Amphibious

Henley target tug of No.631 Squadron at Towyn. Note the coarse nature of the sea-grass of the airfield.

Overshooting at Towyn on 8 July 1944, this B17 Fortress of USAAF was on its way to RAF Valley from North Africa in bad weather when it left its planned course and ran out of fuel.

Training Wing, which was at this time training with DUKWs for the D-Day landings. They sought permission to recover the Henley and on the 21st they brought the aircraft ashore. The Army units needed something to do other than train.

There were in excess of 6,000 servicemen in and around Towyn at the time. Station strength in February 1944 was sixteen officers, twenty-five NCOs and 185 airmen. Flying hours achieved during the month was 334 on Army Co-operation and target towing.

A US Army Air Force B-17 Fortress, on a flight from North Africa to Valley on 8 July 1944, became lost in foul weather with little fuel left. It hugged the coast of Wales until Towyn appeared out of the clag. It was really too small an airfield, but the other options were unattractive, so the landing was made and it was nearly a good one. The bomber overshot across the railway and hit an air-raid shelter, causing a fire. All fifteen on board were unhurt and scrambled clear while the station fire party put out the fire. The aircraft was 42-31321 CCM of the 390th Bomb Group.

At the end of 1944 No.631 Squadron had twenty-one aircraft on charge: eight Henleys, nine Martinets, two Hurricanes, an Oxford and a Tiger Moth hack.

Miles Martinet of No.631 Squadron, towing a drogue over Cardigan Bay.

The centre of the apron between and in front of the two Bellman hangars at Towyn. There is nothing left of the technical site now. (I. Jones)

The concrete base of the eastern Bellman hangar. (I. Jones)

Looking to the north-west from the technical site is one of the best – at 1,200 yards – of the runway options. (I. Jones)

The pilots of the tugs flew 2,000ft from the beach at Tonfanau where the guns were waiting. They would tow at 140mph; any faster would send the engine temps up. There were several types of target, the largest being a red flag 40ft long and 6ft wide, which was towed at various heights.

There were a variety of drogues, too; the planes carried about a dozen on board and were streamed about 400ft behind. There were silent routines – no gunfire, no tow, just gun-laying practice or radar calibration. Another exercise they were involved in was called submarine simulation. In this they towed a white drogue on about 50ft of cable, very low, in fact hedge-hopping, approaching from inland at about 180mph. They aimed to pass over the line of guns on the promenade at right angles, heading out to sea. After twenty-five seconds (approximately 2,000 yards) and at sea-level, they cut the tow and got the hell out of there, as the first shells would be on their way. The drogue, after being cut, would serve as a target on the surface, floating until smashed.

In the early days a Hurricane was used as a gun-laying, radar calibration aircraft. Later there were Spitfire Vbs to simulate Flying Bomb (VIs) attacks. This consisted of approaching from inland in a shallow dive at full throttle, aiming to pass over the guns and out to sea at approximately 2,000ft. That is where the Spitfire 16s came in, because they could approach the speeds of the Flying Bombs in true simulation at 400mph.

The Vultee Vengeance at Towyn was considered excellent as a target tug, unlike some stations where they were disliked and the Martinet preferred, but when one flew the Vengeance it had to be from the runways at Llanbedr. This was not as silly as it seems, because many of the flying aspects of Towyn's schedule did take place from Llanbedr when the weather was poor.

After D-Day the workload fell away and No.631 Squadron's flying hours were down to 280 for the month. Henley L3296 was lost on 15 October 1944 when it made a belly landing and was damaged beyond repair.

Further losses occurred in February 1945 when two of the squadron's aircraft crashed, both piloted by the same man. The first was a Hurricane, which overshot on landing and ran into the railway line resulting in Category E damage. It was flown by P/O H.H. Russell. On 28 February he also piloted a Henley which crashed in the Dysynni Estuary near the airfield. He did not survive this second accident.

Llanbedr had had its two Armament Practice Camps disband and it was decided to permanently transfer No.631 Squadron there from Towyn to make use of the hard runways and facilities at the better station in May 1945. That left Towyn empty before it was taken under the wing of No.22 Group Technical Training Command. Two months later, however, on 25 July 1945, Towyn was closed. It was transferred to the War Office and became an army camp, used by territorials on summer camps. Post-war 'Z' reservists used eight tented camps at Morfa Towyn. Then, during conscription, it was used to train conscripts up until 1960. It later became an Outward Bound Adventure Camp for women and the landing ground was turned into a sports field.

Tonfanau Camp was reopened in 1974 to house approximately 1,000 Asians that had been forced to leave Africa. The Army used the camp to house a mountain climbing school for a period after that. The few buildings that remain are accommodation huts and ablutions. The dispersed sites, No.1 for airmen and No.2 for WAAF, have been demolished.

At the current time a Mr John Holland has purchased the Morfa Towyn airfield site and is to develop it into a leisure facility when the appropriate planning has been approved.

The Airfield

The airfield consisted of two Bellman hangars, two Blister hangars and two Bessoneau canvas hangars, though none extant.

The grass runways were: east–west, 700 yards; north-east–south-west, 1,100 yards; north-west–south-east, 1,200 yards. The control tower was to the specifications of drawing No.952/40.

There were the customary technical site buildings, including a Dome Trainer (drawing No.73/42). None extant.

In 2002 the expanse of grass that was once neat and maintained was a wild field of clumped turf, but I have since discovered that ploughing and re-seeding has resulted in a first-class reproduction of the days when the station first opened.

The airfield at Llanbedr in its original form on 15 April 1942, when built as a fighter station and with just the two runways. There were blister hangars around the perimeter track and very little else. (Welsh Assembly Government Photographic Archive)

TWO
RAF Llanbedr

Grid reference SH 570260, on the A496 in Gwynedd

Constructed during mid-1940 to 1941 in a coastal area with sand dunes on three sides of the airfield and the small town of Llanbedr and the mountains on the other – eastward – side, RAF Llanbedr opened on 15 June 1941. It was intended as a forward base for fighter squadrons defending the Irish Sea area and was controlled by RAF Valley as part of No.9 Group.

The opening up party arrived on 23 July 1941 to find that their new home was set in grand scenery, with the sandy beaches and the sea on their doorstep, Harlech Castle up on its crag to the north and the Mawddach Estuary to the south at Barmouth.

The first aircraft to arrive were not the ones planned or intended, but a detachment of six Ansons of the No.6 Air Observer and Navigation School from Staverton. On 20 September one of these aircraft, with five men on board, was returning to Llanbedr from Worthy Down when it became lost. It crashed into the sea near Bardsey Island, disappearing without trace, though the pilot's body was eventually recovered from the sea near RAF Bodorgan a month later.

The Spitfires that had been intended did arrive on 4 October 1941. The unit was No.74 Squadron with Spitfire IIs from Acklington, under the command of Sqn Ldr P. Ridley DFC.

The squadron's task was to protect the convoys sailing the Irish Sea and to intercept raiders heading for Merseyside. Soon an opportunity arose for combat when, on 13 October, aircraft were scrambled to catch raiders over St George's Channel, but no contact was made. The first combat was on 27 November.

Two aircraft were scrambled on this occasion; P/O Williams and Sgt Ingle were sent to intercept three JU 88s over Cardigan Bay. After making contact with the enemy aircraft off St Davids Head, P/O Williams attacked one from astern and caused enough damage to justly claim a probable. He then gave chase to the other two aircraft, but his Spitfire was hit by return fire from the JU 88s and dived into the sea. Williams's body was washed up on the beach at Pembroke on 17 December. The surviving pilot also made a claim for one probable hit.

The squadron lost another pilot a few days later when Sgt Brown crashed into the sea and a further one the day before Christmas, after an accident while carrying out formation flying.

The squadron was posted to Long Kesh, Northern Ireland, on 13 January 1942. Two days later five officers and 130 airmen left by rail and two Harrow transports landed at Llanbedr to take the rest of the squadron to their new station. Thirteen Spitfires flew with them. They were replaced by No.131 Squadron from Atcham on 8 February, under Sqn Ldr Pedley, equipped with Spitfire Vs. The month gap between No.74 Squadron's departure and the arrival of No.131 Squadron was to enable Drem lighting to be installed at Llanbedr.

This squadron was moved to Valley on 3 March, but was forced to return to Llanbedr in mid-April because of the problems that existed at the time at Valley. They were plagued by sand blowing all over the airfield, affecting the serviceability of the aircraft. There was also a snag with communications; the RT quality was affected by the intervening high ground between Valley and Llanbedr. In May, after completing exercises with the Army, they were posted to Merston.

Avro Anson Mk I of No.6 Air Observer and Navigation School at Llanbedr in August 1941.

Spitfires of No.131 Squadron lined up for inspection. They were at Llanbedr from February to May 1942 along with Spitfire Vs.

The next fighter squadron was No.232, arriving from Atcham on 4 May with their Spitfire Vs. Most of their time was spent working up as a unit. Irish Sea patrols were also flown, but they were being prepared for the Dieppe raid, although they were unaware of this at the time. They were detached to Merston for the raid in late June before returning to Llanbedr and picking up the routine until leaving for Turnhouse in August.

No.41 Squadron was next to arrive, bringing their Spitfire Vs, on 9 August 1942. They had an intercept on the 14th, when Yellow section was ordered to catch a JU 88 east of Dublin. They did this and hits were observed, but the raider managed to escape over neutral Eire. On 22 October 1942, three Spitfires of 'A' Flight, when on formation flying practice in cloud, 6 miles east of Towyn, flew into Hendre Mountain killing all three crew and, of course, writing off the aircraft.

During this period of late summer of 1942, Llanbedr was beginning to be used by many small units, mostly as lodgers, but staying for a long time. The USAAF brought their Lightnings over from Atcham in short detachments for operational training, including air-to-air firing. These were followed by eight P38s of the 48th Fighter Squadron in late August and sixteen of the 49th Squadron in mid-September. Two of these aircraft were ordered to intercept a raider near Carnsore Point, Eire, on 26 September but failed to make contact. In the meantime, another Lightning was lost when it crashed into the sea off Harlech on the 27th.

A target-towing service was provided for these US units by the Lysanders of the 2025 Gunnery Flight of the USAAF, which had an establishment of twelve Lysander IIs. They also opened several new air-to-ground firing ranges at Tal-y-Bont beach at Lake Trawsfynydd and an anchored buoy off Shell Island, to the west of Llanbedr Airfield, for bombing practice.

At the same time, the Royal Navy put a detachment of No.776 Squadron at Llanbedr, from Speke, and became firmly established, staying for two years before moving to nearby Mona in

Right: Llanbedr in 1960, showing the development needed by the Rocket Propulsion Establishment. Note the new long runway and the hangars with aircraft servicing platform in front, and also the extensive area of sand dunes to the west where, during the war, the target ranges were situated. (Welsh Assembly Government Photographic Archive)

Centre: Spitfire Vs of No. 41 Squadron served at Llanbedr from 9 August 1942 until February 1943.

Bottom: This Bellman was known as the Meteor hangar by those on site and is building No.2 on the site plan. (I. Jones)

April 1944. This unit, again, was in the role of target-towing and was equipped with approximately six Skuas. The unit was to provide targets for elements of the Royal Navy in Cardigan Bay.

On 25 February No.41 Squadron, the airmen of which were bored by the lack of action, was posted to High Ercal, and on their leaving became the last operational fighter squadron to be resident here. From now on it was to be primarily an Armament Practice Camp.

The next unit to arrive was No.1486 Fighter Gunnery Flight which had been pushed out of Valley. This was because of the decision to make that station a massive terminal for transatlantic deliveries of the four-engined bombers and transport types that were beginning to pour in from the States. The scheme was named 'Transat'. There was no room for anything else at Valley. The gunnery unit arrived on 1 July 1943; on 15 October it was disbanded and redesignated No.12 Armament Practice Camp (APC), with six Martinets, two Masters and three Lysanders. The following month saw the arrival of No.13 APC – formed at Weston Zoyland from No.1492 Target-Towing Flight – bringing Masters, Martinets and Hurricane IIcs.

Both of these APCs disbanded on 21 February 1945 at Llanbedr. During their time there they were in No.12 Group Fighter Command.

By now the station was busy with the various bombing and gunnery flights. The US Lightnings had been replaced by the Thunderbolt and the American Air Force was active in the

This was the original RAF watch office/control tower. Under DERA it became the Flying Operations, building No.27 (drawing No.343/43). (I. Jones)

This was T2 hangar East with an annexe, building Nos 265 and 266. (I. Jones)

Nos 12 and 13 Armament Practice Camps used the Miles Master trainer, such as this one at Llanbedr.

area. Bombers on liaison flights were occasional visitors and a Liberator that had flown all the way up from Marrakesh missed Valley, its destination, and landed at Llanbedr in desperation.

A Wellington experiencing severe icing conditions at 15,000ft over the nearby mountains luckily managed to land at Llanbedr. On the following day their luck ran out when, a few moments after taking off, it crashed a half mile from the airfield, killing one of the crew and seriously injuring two others.

From November 1943 until May 1944 a succession of twenty-six British and allied fighter squadrons came to these two Armament Practice Camps on short detachments of ten days or so, depending on the weather. The aircraft of the visiting units were either Spitfires, Mustangs or Typhoons. They made use of the air-to-ground ranges at Dyffryn, next to the airfield, and air-to-air shooting at drogues towed by the Martinets. This was really a crash course to enhance the performance of all the units ready for the D-Day landings in France.

In March 1944 No.129 Squadron arrived here to convert from Spitfires to Mustang IIIs, doing their working up on the ranges.

The squadrons attending the Armament Practice Camps are as follows, in ascending squadron numbers:

Squadron	*Dates*	*Moved to*	*Aircraft*	*Code*
No.66 Squadron	22 February–1 March 1944	Fairwood Common	Spitfire LF 16e	LZ
No.129 Squadron	30 March–3 April 1944	Coolham	Mustang III	DY
No.131 Squadron	9 February–4 March 1942	Merston	Spitfire Vb	NX
No.164 Squadron	12 April–21 April 1944	Thorney Island	Typhoon 1b	FJ
No.168 Squadron	21 January–4 February 1944	North Weald	Mustang 1a	–
No.183 Squadron	11 April–22 April 1944	Thorney Island	Spitfire IX	HF
No.193 Squadron	6 April–11 April 1944	Needs Oar Point	Typhoon Ib	DP
No.195 Squadron	11 April–19 April 1944	Disbanded	Typhoon Ib	–
No.198 Squadron	30 March–6 April and 22–30 April 1944	Thorney Island	Typhoon IB	TP
No.232 Squadron	16 May–3 August 1942	Turnhouse	Spitfire Vb	EF
No.268 Squadron	7–20 February 1944	North Weald	Mustang II	Nill
No.302 Squadron	1–7 March 1944	Northolt	Spitfire IX E	WX

This A35 Vengeance target tug of No.1 Gunnery and Target Towing Flight, USAAF, is being prepared on the ASP between the two T2 hangars. Note the fitter's bicycle against the wheel.

This overhead shows the air-to-ground moving target railway line at Traeth Dy, near Harlech. (Welsh Assembly Government Photographic Archive)

No.306 Squadron	19 December 1943 –1 January 1944	Heston	Spitfire Vb and Mustang III	UZ
No.308 Squadron	8–15 March 1944	Northolt	Spitfire IX E	ZF
No.312 Squadron	2–18 December 1943	Ibsley	Spitfire Vc	DU
No.315 Squadron	19 December 1943 –1 January 1944	Heston	Spitfire Vb	PK
No.322 Squadron	15–30 November 1943	Woodvale	Spitfire Vb	VL
No.322 Squadron	19–23 May 1944	Merston	Spitfire IX	–
No.331 Squadron	5–21 January 1944	North Weald	Spitfire IX	FN
No.340 Squadron	15–19 May 1944	Merston	Spitfire IXb	GW
No.341 Squadron	11–16 May 1944	Merston	Spitfire IXb	NL
No.349 Squadron	6–11 April 1944	Hornchurch	Spitfire LF IX E	GE

No.350 Squadron	8–19 February 1944	Hornchurch	Spitfire IXb	MN
No.485 Squadron	21–28 March 1944	Hornchurch	Spitfire IX E	OU
No.504 Squadron	28 January–4 February 1944	Hornchurch	Spitfire IXb	TM
No.602 Squadron	13–20 March 1944	Detling	Spitfire IX E	LO
No.609 Squadron	22–30 April 1944	Thorney Island	Typhoon 1a	PR

All these units finished their stint at Llanbedr before D-Day, unlike the APCs at Fairwood Common which were to continue to work until the end of the war.

Although the APCs were disbanded on 21 February 1945, the work had died away long before that. The Americans continued using the ranges up to the end of hostilities. In July 1944 a group of ATA pilots were flown in to collect eight Martinets and within a week all the APC's aircraft were gone.

The ranges were at Dyffryn and at Traeth Dy (a light railway moving target, air-to-ground range). A bomb target was anchored off Shell Island, an air-to-ground range was at the beach of Tal-y-Bont, there was a bombing range at Penychain and there was a range at Trawsfynydd, but exactly where and in what form is not known.

The war ended with just the one flying unit at Llanbedr, No.631 Squadron. It had been formed at nearby Towyn by merging Nos 1605 (Target Towing) and 1628 (Anti-Aircraft Co-operation) Flights on 1 December 1943 and was coded CD. Due to the grass airfield at Towyn being flooded frequently, an arrangement with Llanbedr would often see the No.631 Squadron's aircraft using the runways of the larger station. When the airfield at Llanbedr was vacated by the APCs in May 1945, No.613 Squadron moved in completely, with its Vengeance IV as tugs, Hurricane 11cs, Spitfire Vbs, Spitfire 16s, Harvards and, later, Vampire F1s and Martinets. Towyn was closed down as this squadron was its only resident unit. This unit stayed and served the Anti-Aircraft School at Tonfanau and other batteries along the coast. They were still having accidents though. A Martinet came down in the sea off Barmouth and the lifeboat found P/O Davis sitting in a dinghy, but the tow operator was lying face down in the sea, having drowned. The aircraft had been target-towing off the coast at Tonfannau, the anti-aircraft range. The pilot throttled back to deposit the drogue, and when the throttle was reopened there was no response and the Martinet fell in to the sea. A subsequent inquiry, following the recovery of the aircraft, showed that the rod linking the throttle with the carburettor butterflies had not been wire locked at the carburettor end. Disbandment came for No.631 Squadron on 11 February 1949 at Llanbedr, only for them to reform as No.20 Squadron and disband again on 16 October 1951 at Valley.

The work with the artillery continued at Llanbedr with the introduction of a new unit after the disbandment of No.20 Squadron. The No.5 Civilian Anti-Aircraft Co-operation Unit (CAACU) was eventually to be equipped with a complement of the latest and most powerful target tugs and attack simulators – the Mosquito TT 35, the Beaufighter TT X, Vampires and Meteors, which were flown by civilians – and the whole operation was managed by a contractor, in this case Short Bros & Harland of Rochester. This was to continue until 1 January 1955, when the unit transferred to Woodvale where sixteen years later, on 30 September 1971, it was deleted completely.

By 1951 the Royal Aircraft Establishment at Farnborough (later the Royal Aerospace Establishment) had formed a Rocket Propulsion Department and in 1958 redesignated it the Rocket Propulsion Establishment, using facilities at Llanbedr in 1954 and at West Freugh in 1956. As Llanbedr was in an area north of the testing ranges used by the now missile testing at Aberporth, overlooking the same Cardigan Bay, and as it was realised that unmanned targets were needed for the new weapons coming on stream, Llanbedr was chosen as the new Target Development airfield.

A new long runway of 7,500ft was built running roughly north–south. Two new T2 hangars were erected with workshops and laboratories in the annexes, but most importantly of all was the electronic equipment. This would control the unmanned target aircraft into the air, then to its destination, bring it back and land it, without sacrificing it. The target had to be fast and controllable at any altitude that was required by Aberporth's Missile Testing Unit, and had to be capable of towing a drogue target.

The first of these were surplus Fairey Fireflies; when the radio guidance equipment was installed in these aircraft they were known as drones. As well as the drones, they also needed

Parachute store and drying room at RAF Llanbedr in 2002. (I. Jones)

T2 hangar west and annexe, building Nos 269 and 270 on site plan. The two hangars and annexe are separated by the service road that leads to the ASP via a security gate. (I. Jones)

Spitfire Mk 16E of No.631 Squadron at Llanbedr. This aircraft was used for fast passes over Tonfanau Anti-Aircraft School and gun-laying practice.

This is another Spiffire Mk 16, also at Llanbedr but this time serving with No.20 Squadron, coded 6D. This change of squadron number was simply a designation, recognising the fast reduction of units of the RAF post-war.

The Fairy Firefly as a retired carrier-born type that became the ideal vehicle for the radar and other electronics needed for chasing the Jindivik.

aircraft to fly with the target, to observe the attitude and behaviour of the drone. These were also Fireflies. The pilot would be in touch with the ground controllers and have a measure of control himself by wireless. These were called Shepherds.

The first unmanned air vehicle (UAV) flight was a Firefly in 1952. Since then the Firefly has been used extensively and the conversions kept Faireys busy. Since then, the Firefly has been succeeded by the Meteor and the Jindivik, and is now complemented by the air-launched Stiletto supersonic target.

After RAE management the airfield was put out to private contractors, and the first of these was Short Bros. Since then it has been Airwork, FRA Serco to Serco operations, and under the auspices of DERA (Defence Evaluation Research Agency) with West Freugh and Benbecula as other ranges.

95 per cent of the work is for the MOD but the weapon manufacturers can hire time on the range. Other capabilities at Llanbedr include supplying manned aircraft for tracking requirements

and providing towed and released targets for Off Range users such as the Royal Navy and MOD Project Offices.

At this time in the early 1950s No.5 CAACU was still operating from Llanbedr. The main work was at Tonfanneu, the anti-aircraft school just north of Towyn. The school had heavy and light artillery. The two main exercises were Z6 at 6,000ft for the heavy guns and X1.5 for the light guns at 1,500ft. The runs could be live firing at a towed sleeve or silent (no firing) in a dumbell pattern with 360-degree turns at each end of the runs along in front of the guns. The sleeves could be ordinary, or radar type, the sleeve having a wire mesh. The aircraft was usually the Beaufighter TT10 and later the Mosquito TT35, from March 1954 on. For the silent runs it was mainly Spitfire 16s or Vampire 3s and 5s. In April 1954 the Beaufighters were retired and the Spitfires went in September 1954. Meteor 8s arrived in August 1955.

The Ty Croes Range was a trials establishment where all types of anti-aircraft guns were tested, some being radar controlled. At Llanbedr it was still under the RAF and RAE in the 1960s and '70s, with the target development unit flying Hunters, Meteor 7s, 16s and TT 20s, a Devon and Ansons for hacks. In 1960 it was under the Ministry of Technology, later the MOD (Procurement Executive), and the first flight of the Jindivik was made. This UAV was of UK design and early production was manufactured in Australia in a mutual agreement with the UK. It has been used as

A store of Fireflys in one of the T2 hangars at Llanbedr.

The Gloster Meteor took over the shepherding from the Firefly to use the extra speed of the jet.

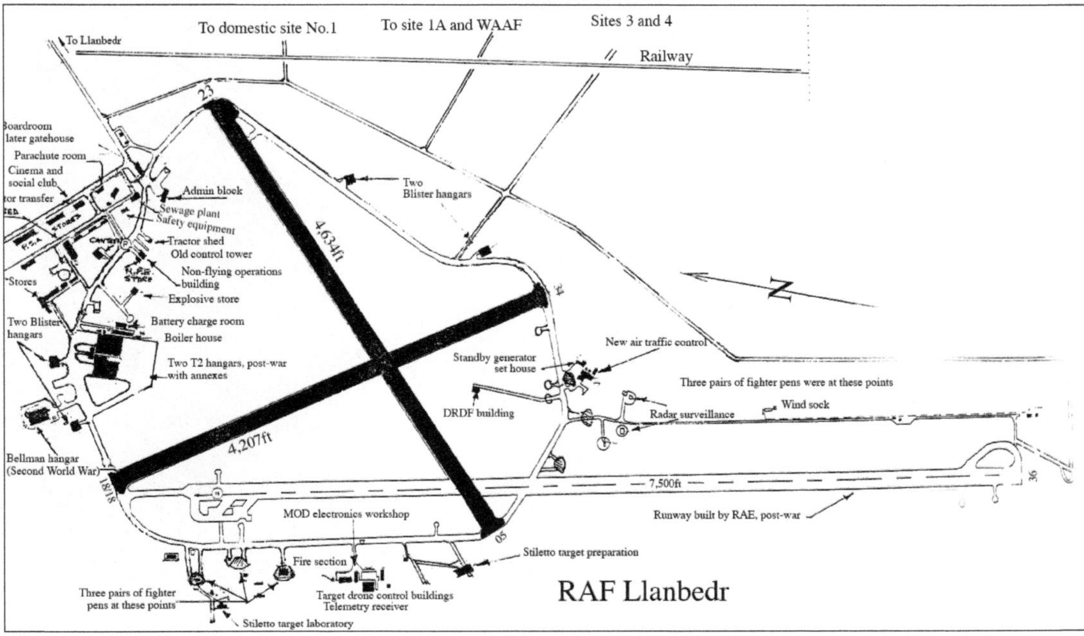

This map shows the airfield at Llanbedr at its peak, with new air traffic control, radar for the missile firings and the super runway. (Copied from DERA drawings by I. Jones)

a direct target but usually as a tower of targets, to save them from destruction not only for the cost factor but because the Jindivik is no longer in production and there are only a few left. By the year 2000 there had been 7,000 Jindivik flights, the attrition rate being one airframe per year over the last two years, though the rate was much worse in the early years, 236 having being lost in total.

Currently the base is flying the Sea Vixen, Hawk and Alpha-Jet as chasers and shepherds. Two Canberras are used for the release of the supersonic Stiletto target and for towing targets such as Rushton Tow and the Low-Level Height Keeper, as an Exocet simulator.

Jindivik-type operations are to be co-located to the Hebrides Range to fulfil future trial requirements.

The Airfield

The land where the airfield sits was, and still is, to the locals anyway, known as Morfa Dyffryn. At the end of the war there were just the two runways: north-east–south-west, 23/05, 4,634ft, and north-north-west–south-south-east, 34/16, 4,207ft.

The long runway, built to accommodate the pilotless drones, was to the requirement of the Royal Aerospace Establishment and was constructed in about 1955. This superb runway runs north–south and is 7,500ft long. Like the other two runways it is 150ft wide, of concrete and tarmac.

The wartime station had one Bellman and four Blister hangars. There were six single-engine type pens that sheltered two aircraft in each.

All the domestic sites were over to the east of the field on the other side of both the main road and the railway. Two new T2 hangars were built for RAE at the same period as the runway. All the larger hangars have annexes adjoining for offices and, of course, there are other offices, workshops, laboratories and stores. Dangerous chemicals and fuels are along the length of this runway. The remote controls for these unmanned air vehicles are situated at the runway ends.

Unfortunately the news in 2003 is that both Aberporth and Llanbedr, together with the other ranges, have been privatised by a consortium that absorbed DERA, operating under the name of QinetiQ. Although retaining Aberporth, it had no use for Llanbedr after the summer of 2004. Hopefully something can be done to utilise this base with its fine runway and wonderful staff.

The Jindivik; like all the others it is now 'retired'.

Ansons T22, left, and T21, right, with Llanbedr's Communications Flight.

Jindivik with Meteor shepherd, over Cardigan Bay.

Jindivik Gate Guard at Llanbedr in 2002. Note the old control tower, second building on the left. (I. Jones)

Canberra target-tower in T2 hangar west. (I. Jones)

This view shows the new control tower and the elaborate radar and communications complex on the southern perimeter track. (I. Jones)

It is 26 May 1961 and the Butlins Holiday Camp is active. Broomhall Aerodrome is at the top of photograph. (Welsh Assembly Government Photographic Archive)

THREE

The Aerodrome at Broomhall

Grid Reference SH 410370, near Pwllheli, Gwynedd

Broomhall Estate with its manor house was the home of Colonel O.J. Lloyd Evans, the squire of this Lleyn peninsular district, and upon his death his son William Prys Evans inherited the house and estate.

William Evans was a keen flyer, having obtained his Private Pilot's Licence in 1932 at the age of twenty-seven. He purchased a DH Hornet Moth G-ADJY, built a hangar for it, then cleared and prepared an airstrip to the north of the house, from which he made many flights, and others – friends and business people – also flew in.

In 1934 he was primarily responsible for the formation of gliding in Gwynedd, along with a group of enthusiasts. He made his own airfield available for the primary training of what was only slowly to become the Pwllheli and District Gliding Club.

On 6 April 1937 Mr Evans, flying his Hornet Moth, took off to visit his brother in Surrey. For some reason failed to gain height and crashed into a high estate boundary wall, wrecking the aircraft and killing Mr Evans.

The estate then passed to another son who had no interest in either the estate or flying and parts of the land were sold off. No flying took place here until the end of the Second World War.

After the Second World War it was decided to sell off the whole estate and in October 1946 this was achieved. The buyer was none other than Mr Billy Butlin of the holiday camp fame, who was about to open his latest camp just a mile to the south-east of Broomhall. He wanted the estate to grow and then supply produce from the estate farm to his holidaymakers, and to use the airfield to entertain his guests with aerobatics and cheap air trips. Furthermore, at the back of his mind, the idea formed that perhaps he could develop the flying to include bringing some of his holiday guests in by air and return them home at the end of their holiday to their local airport.

Built by Billy Butlin, the holiday camp opened in 1947, having been refurbished and altered to meet its new role as it had originally been built to accommodate the sailors of HMS *Glendower*, a Royal Navy Shore Establishment.

The airfield was extended after the purchase of more land from Rhedynog Farm and it was now possible for two runways to be used: north–south, 1,650ft (later extended to 2,150ft) and east–west 1,460ft (later extended to 1,800ft). Another much larger hangar was erected near the hangar from the 1930s built by Mr Evans; a wind sock and a control caravan were situated out near the runway crossing.

The actual house of Broomhall was occupied by the Butlins Camp Controller, Major Frank R. Bond MBE. The estate came to be managed under the name Broomhall Enterprises.

A number of aircraft were bought, the first, in July 1947, being Southern Martlet G-AAYX, which operated from the estate giving aerobatic displays. It was sold on in December 1949. Aircraft were bought to provide trips to the holidaymakers and also for communications with Butlins' other camps and its headquarters in London. Percival Proctor G-AGTE was purchased in March 1946, followed in November by Airspeed Consul G-AIKU, four Fairchild Argus in 1947

The fatal crash of Colonel Evans's wealthy son on his private airfield at Broomhall. These are the remains of Hornet Moth G-ADJY, on 6 April 1937. (Ioan Gwylim Jones)

The hangar built in the 1930s by William Prys Evans. It would just accommodate a Rapide, or at least so it is said. It stores agricultural goods now. (Ioan Gwylim Jones)

– G-AJNN, G-AJGW, G-AGPE and G-AJVI. Three Dragon-Rapides – G-AKTD, G-AKOG and G-AKMD – arrived in 1948. This was to prove as good as it was to get, and a decline in numbers followed as Butlin Aviation, as it was by now called, began to outsource its work.

Cambrian Air Services in Cardiff had been providing flights at Broomhall since 1947, with Auster Autocrat G-AIGY for the holidaymakers, and was to provide the only post-war fatalities here when, on 1 August 1950, the aircraft crashed while beating up the airfield, killing the pilot, Capt. Veale, and his passenger.

A second Autocrat, G-AFWN sent from Cambrian, took its place. The only other accident was that of Percival Q6 G-AFIX belonging to Starways of Speke, which came to grief at Broomhall without injuries to the pilot on 6 May 1949.

After negotiations with Butlins, Cambrian took over the operation of the airfield, as they were to do at Withybush (Haverfordwest), Fairwood Common (Swansea) and Pengam Moors (Cardiff). The transportation of passengers from Cardiff to Broomhall and back again was begun.

The Southern Martlet G-AAYX, purchased by Butlins on July 1947 for joy rides and aerobatics. It was sold in December 1949. (Courtesy of Cliff Knox)

The DH 89 Dragon Rapide G-ALAT of Cambrian Air Services, seen here at Cardiff Pengam Moors in 1950, parked between the two Bellman hangars. This particular aircraft was the most involved of the fleet in the flying from Broomhall. (Captain Ken Wakefield)

Later it became Cardiff to Broomhall Butlins and on to Dublin Butlins. For full loads of seven people the fares were £3 15*s* return to Pwllheli and £8 5*s* to Dublin; the route was operated with Rapides from Cardiff.

Other companies worked their aircraft to Broomhall, such as Airmotive to and from Liverpool with one Rapide G-AFOI, but this ceased in June 1951. The concession at Broomhall was taken up by an ex-Imperial Airways pilot, Capt. Morris Guinane, in 1953. He planned to start operating at Easter 1953 under the name Dragon Airways, but the Ministry of Civil Aviation was not satisfied with the airfield or Capt. Guinane's licence to fly commercial aircraft. During May 1953 obstacles on the aerodrome's runways were to be removed and improvements made. Capt. Guinane had to go on a refresher course, so flying did not start at Broomhall's summer season until June. Then, after a month of operating the service, a nasty accident occurred. Rapide G-AIYP, piloted by Guinane, took off with seven passengers. Near the hamlet of Pentreuchaf the aircraft encountered a thick blanket of mist over rising ground and Guinane decided to force-

This Rapide of Dragon Airways, G-AIIPT, operated between Speke Airport and Broomhall. It is seen here at Speke. (Courtesy of Mike Lewis)

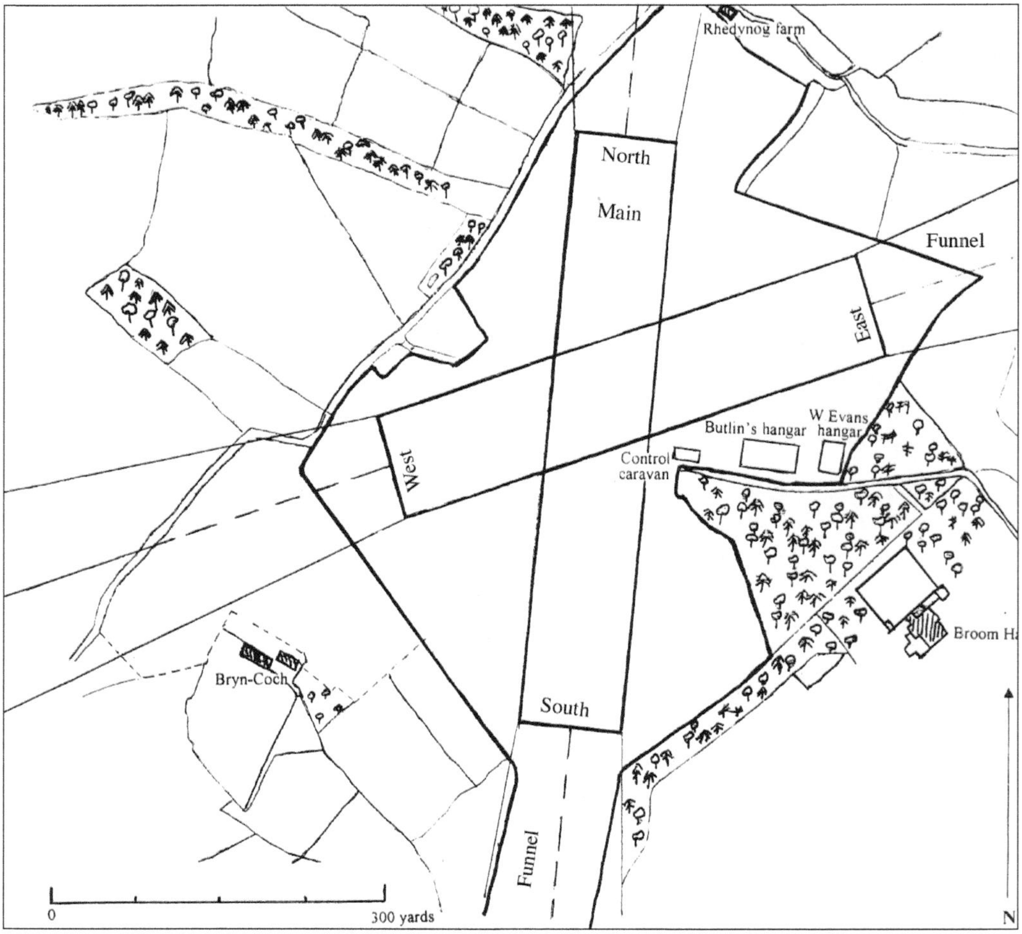

Broomhall Aerodrome post-war. (Drawing courtesy of Roy Sloan)

Lewis Plumstead, the pilot of Rapide G-AJCL, at Broomhall in the summer of 1949. (Lewis Plumstead, via Ioan Gwylim Jones)

Dragon Rapide G-ADDI of Chrisair operated in and out of Broomhall, but is seen here at Luton clearly displaying its livery.

The Dornier DO 28A G-ASUR was one of many private aircraft based here during the 1960s. (Ioan Gwylim Jones)

land in a cornfield, but ran into a hedge. Moments later the aircraft burst into flames; the slight delay in between the collision and the conflagration allowed the occupants to escape unharmed, but the aircraft was destroyed.

It had been Capt. Guinane's intention to run scheduled services from Broomhall, but his refusal to pay his share of the costs involved in preparing the airfield and financial loss, for one reason or another, led to the failure to provide flights for the Butlins holidaymakers. A state of animosity existed between the two companies.

November 1955 saw the company Dragon Airways acquiring new routes and move its base to Newcastle, quitting Broomhall and Butlins.

Another regular at Broomhall, the Jodel DR250-160 G-ATTM. Note the end wall of the old hangar at the rear. (Ioan Gwylim Jones)

August 1958 saw the airfield play host to three Airspeed Consuls disguised as Japanese bombers for the shooting of the film *The Inn of the Sixth Happiness*, which was partly shot at Beddgelert in Snowdonia, and there were the daily comings and goings of privately owned light aircraft to this lovely spot.

After the departure of Dragon Airways, the concession to provide pleasure flying was taken up by Mr Vivian Bellamy of the Hampshire Aeroplane Club using Auster Alpha G-AGYT. In 1958 this contract was signed and it continued to fly this task until 1967 when, oddly, flying of all kinds was diminishing at Broomhall. Perhaps the advent of the family motor car had something to do with this.

The next company to take up the concession was Chrisair. They used DH Dragon G-ADDY in the early 1960s with little profit and in 1967 Billy Butlin stated that 'the airfield was of little advantage as an airfield and could be returned to the farm when the present contract runs out'.

Broomhall was to become part of Gwynedd's aviation history. By 1968 the main hangar had been demolished, though there was some flying activity with the presence of a Cessna and a Gemini here as late as 1974. There were microlight rallies held here during the 1990s on land belonging to Rhedynog Farm, the same land that was sold to Butlins in the late 1940s to extend the runways and now sold back.

Mr Ioan Gwylim Jones was a lad on the Rhedynog Farm throughout the years of aviation and the aerodrome was virtually in his backyard. Ioan met many aviators and filmed much of what went on here. He alone went to the trouble of listing every aircraft that visited Broomhall.

Mr Roy Sloan has written a history of Broomhall's aerodrome, in his impeccable manner, the only history that I know of this place and I thank Mr Sloan for allowing me to use some of his work that was published in *Rapide* magazine. However, we both relied on Ioan Gwylim Jones.

The identity of the type of the larger hangar at Broomhall is unknown as no photo exists. There was a control caravan for the airfield out on the field near the intersection of the runways (see the overhead photograph on page 34).

On a visit here in June 2007 I had the benefit of meeting Mr Ioan Jones and his wife, Sally. Ioan is retired now and his son has taken on the farm at Rhedynog, whose fields formed a part of the area of Mr Butlin's aerodrome. The new owner of the Broomhall Estate, Nigel Coatnan, was contacted to see if entry to his part of the aerodrome was possible to take photographs. He agreed and also joined us in reliving those years of activity here. I say 'relive' because Ioan is a fount of knowledge and has an excellent memory. I took many pictures but space limits how many can be shown here – only those illustrating the runway ends appear. This is a very lovely place that has tranquility, and an unchanging history.

Above: This shot was taken from the corner of the field at the north end of the north–south runway, at the entrance to Rhedynog Farm. Across the field is the funnel extension at the south end. This was the longer of the two runways. (I. Jones)

Right: A clearer view of the south extension funnel. The ground appears to have become lumpy since flying ceased here. (I. Jones)

Looking into the funnel-like extension from the eastern end of the west–east runway. The long shed is a recent acquisition and played no part in the aerodrome's activities, but the smaller shed to its right did – it is the old hangar. (I. Jones)

The western end of what was the west–east runway. The fence is a new one that divides the one field up into several that were returned to their original farms. Over the brow of the rise that forms the horizon is a further flat area of land serving the landing ground. (I. Jones)

This RAF photographic reconnaissance unit overhead shows Penrhos Airfield as it was on 10 July 1946. Most of the hutted camp on the north-west corner was by then occupied by the Poles. (Welsh Assembly Government Photographic Archive)

FOUR
RAF Penrhos

Grid reference SH 335335, on the A499 3 miles west of Pwllheli, Gwynedd

The station of Penrhos was due to open in January 1937, but actually began its life on 1 February 1937 as No.5 Armament Training Station (ATS). At that time there were other ATSs at Catfoss (Yorkshire), Aldergrove (Northern Ireland), Watchet (Somerset) and Sutton Bridge (Lincolnshire). All these camps were providing visiting squadrons with the facilities to develop their bombing and air firing skills by having in place the targets and safe ranges away from the general public.

On 3 April 1937 thirty Hawker Harts and Audaxes plus five Gloster Gauntlet fighters flew in to Penrhos for a month's armament practice. Acting P/O 'Pat' Pattle led this formation. He was a South African born of English parents who was soon to be posted to No.80 Squadron, flying Gladiators in the Middle East, and was destined to be the top-scoring British and Commonwealth fighter pilot with forty plus victories.

On charge at Penrhos were Westland Wallaces, Mks I and II, as target tugs and Avro Tutors. The unit was disbanded just over a year later, on 1 April 1938, to be reformed as No.5 Armament Training Station, with Wallace IIs, Henley IIIs and Magisters. Most of the detachments that were sent to Penrhos were from flying training schools, such as Tern Hill and Sealand. This unit was to disband on 18 September 1939 at Jurby to become No.5 Air Observer School.

Units attended these camps for a period of one month, practicing firing at 10ft sq. targets on the ground with their Vickers and Lewis guns, which were the standard gun types in 1937, or firing at drogues over the sea in quarter attacks for the front guns and beam attacks for the rear guns. There was high and low bombing and also dive bombing at targets floating off shore, with 8.5lb practice bombs. All of this took place at Penrhos and the target range at Hell's Mouth, some 8 miles south of the parent station.

RAF Penrhos had been built in the heartland of Welsh Nationalism on the Lleyn Peninsular and from the beginning there had been protests and demonstrations by some people at the compulsory purchase of several farms, not only for Penrhos but for the range and airfield at Hell's Mouth. The protests culminated in an act of arson on the building contractor's property by three men who eventually gave themselves up to the police and, after a controversial trial, they served short sentences. It should be remembered that when the crime was committed in 1936 war was not definitely on the agenda.

As the flying training schools came to Penrhos for their turn at air firing (mostly with biplanes), the possibility of war increased and by the outbreak of the Second World War on 3 September 1939 there was a great need for trained aircrews. On the same day as the declaration of war a new unit was formed at Penrhos, No.9 Air Observers School. This unit had been the result of disbanding and redesignating No.1 Air Observers School. Six days later, on 9 September 1939, No.9 Air Observers School opened at Penrhos and the courses began. The school was organised into four flights: 'A' had the target towing with Wallaces and Henleys; 'B' had Battles for bombing practice; gunnery training was under 'C' and 'D' flights, flying Hawker Demons and Handley-Page Harrows. In December they were joined by 'C' Flight of No.1 Anti-Aircraft Co-

The link trainer building (No.72). It is now a shop for the caravan site. (I. Jones)

Westland Wallaces were used in large numbers here as target-towers with No.5 Armament Training Station (ATS) in 1938–39. (Frank Jones)

Hawker Henleys replaced the Wallaces as target-towers at Penrhos No.5 ATS.

A pre-war shot of Penrhos while building was still going on. (I. Jones)

Pillbox/Battle headquarters, on the hill to the north of butted camp. (I. Jones)

The Hawker Demon was in daily use at Penrhos for air gunnery practice, firing at the drogues that were towed by the Wallaces with No.5 Armament Practice Camp. (Frank Jones)

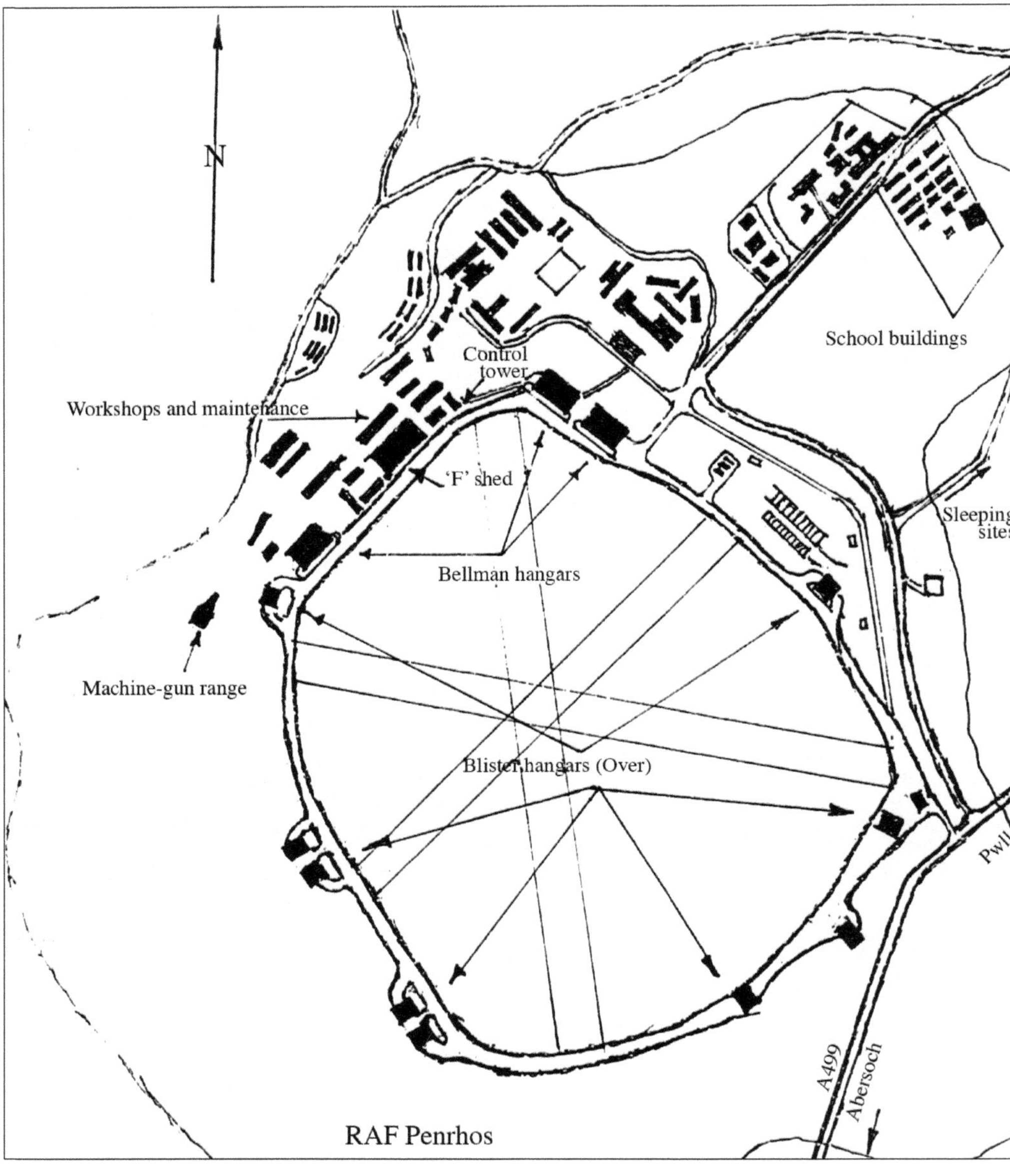

Map of RAF Penrhos illustrating the grass runway options and the buildings on site. (I. Jones)

operation Unit, operating Henleys, and in February 1940 by 'J' Flight of No.1 AACU, also with Henleys. These were flying to assist the artillery at Tonfannau across the bay, until Towyn and later Llanbedr could be brought into being.

The title of No.9 Air Observers School was changed to No.9 Bombing and Gunnery School in No.25 Group on 1 November 1939.

The Penrhos Airfield was roughly circular in plan and was of grass throughout its existence, really only fit for the small biplanes of the 1930s, with a 100ft ridge to the north-west to contend with and a nasty drop over the lip on the southern and western perimeters. Many Blenheims and Ansons came to grief, as the longest run at take-off was a mere 2,300ft. There were many people at RAF headquarters who considered the airfield to be too dangerous and unsuitable to be used for training. Apart from Hell's Mouth Airfield and ranges, another

The gunner in this Bristol Blenheim Mk I is set up for his shot at the drogue, No.9 Bombing & Gunnery School (B&GS) at Penrhos. (RAF Museum)

Penrhos in 1939/40. Blenheim Mk Is and Ansons of No.9 B&GS are parked along the east side of the airfield. (Via E. Doylerush)

outstation under Penrhos was the Marine Section at Pwllheli. No.51 Air Sea Rescue Base was established in Pwllheli Harbour with five locally built patrol pinnaces. They were to maintain the floating targets in Hell's Mouth Bay and to keep shipping away from the area when the bombing range was in use. Two of the boats were armoured as target boats for smoke bombs. Live bombs were only to be used outside the 3-mile limit under stringent precautions and a safety officer would always be on duty in a tower on the beach at Hell's Mouth to oversee proceedings.

In the last months of peace the training of aircrew was hastened and the small airfield at Hell's Mouth was considered to have potential as a relief airfield for Penrhos. Plans were drawn for an extension to the airfield's limits and for three Bellman hangars plus a moving target range to be erected.

In the early weeks after the war began, a sequence of strange events occurred off the shore near Hell's Mouth. On 23 October the duty pilot took off early in the morning and, soon after, spotted a meeting between a submarine and a fishing boat offshore. The submarine was submerging and proceeding in a southerly direction; the fishing boat was given a look-over by the duty pilot, who was convinced that something had passed between the submarine and boat. The number of the boat was passed on to the police and the HQ of Coastal Command. It is important to remember that at the time there was a suspicion in Britain that some people and politicians in Eire might be sympathetic to the German cause and give aid to U-boats. Further suspicious activity was seen on the 26th. At roughly 2 a.m. two white lights were observed at sea off Llanbedrog Head by the guard at the entrance to the station. The local coastguard investigated and it was discovered that a ship had approached St Tudwals Islands and moored there, later moving to Llanbedrog Head. No details of the ship were known and the Senior Naval Officer (SNO) at Liverpool was informed by telephone. He wanted further particulars. Accordingly a Battle took off at 7.30 a.m. and located the ship. The pilot could not make out the name of the vessel but said that it was a small cargo ship at anchor. At about 9.15 a.m. it moved into Abersoch Bay and the following details were obtained by the coastguard: it was the *Eden Vale* from Wexford, flying the Irish flag. The SNO at Liverpool said he would take the necessary action. This episode resulted in a decision to have one Battle in an armed state and at full readiness in case of U-boat activity.

Nothing more was seen, however, until 1 April 1940 when the local coastguard once again reported a U-boat in the vicinity. The duty Battle took off to investigate but found nothing. The following day one of the staff pilots on a training flight spotted a U-boat, which was probably the same one that was spotted the day earlier. Again the Battle took off to search, but failed to find the elusive submarine. There were no more sightings for the duration of the war.

While based at Benson between 17 September 1939 and 6 April 1940, No.63 Squadron had a detachment at Penrhos equipped with Battles and Anson 1s. No.90 Squadron also served at this station briefly from 13 August to 27 August 1939 with Blenheim IVs.

The Demons of the observer school were replaced by Whitleys in the summer of 1940. A large number of Polish officers and NCOs also arrived to take up flying duties and assist in the training of Polish aircrews. There was a preponderance of Polish airmen at the training schools in West Wales, especially at target towing.

On 1 July 1940, a Dornier 17 carried out a bombing attack on Penrhos and machine-gunned parked aircraft. Two officers were killed, two Henleys of No.1 AACU were destroyed and three blocks of officers' quarters were demolished. A hanger was also damaged. Following this the airfield was hurriedly camouflaged and lines painted across the field. Roofs and roads were toned down.

The first of many accidents that were to play a part in the formation of the Mountain Rescue Team occurred on 8 April 1940 when a Blenheim L9093 from No.13 OTU (Operational Training Unit) at Bicester was reported missing. The next day the aircraft was found wrecked on the summit of Carnedd Llewelyn at 3,000ft. The Blenheim had flown into the side of the mountain while in level flight and the crew of four were killed.

Just as the air raid of 1 July was being regarded as a one-off, three months later, on 2 October, an unusually extensive series of raids on this small station began. The German attempt to wipe out Penrhos was seen by some as the result of the worldwide publicity generated by the pre-war Welsh Nationalism and its arson attacks. This attack of 2 October was carried out at dawn by another Dornier DO17, which came in at 200ft and repeated the bombing and machine-gunning of the first raid. Five airmen were injured, one seriously, some contractors' buildings were demolished, the Air Ministry Works Office was damaged and several motor vehicles damaged.

The next day another lone aircraft – this time at dusk – flew in and dropped a stick of bombs just south of Tudweiliog. It was presumed that the attack was meant for the radar station at Nefyn, not the station at Penrhos. On the third successive day there was another attack at first light. Three Dornier 17s swooped over the airfield, dropping HE bombs and incendiaries. There were no casualties, but the landing ground was cratered, the offices of OC Flying and 'A', 'B' and 'C' Flights were demolished. The main hangar – the 'F' shed – was damaged and so were the

Right: One of a series of No.312 Squadron photos taken at Penrhos by unit members and sent to David J. Smith. The historian, who is well-connected to the Czech Veterans Group, has allowed me to use some of them. They come via Zdenek Hurt. Hawker Hurricane IV7066 DU-T, in this photograph, was shot down by a BF 110. It was flown by Flt Lt D.L. Dawborn who, in company with Sgt T. Truhman, had chased the enemy out over the sea on 10 April 1941. This was a new machine and was delivered to No.312 Squadron on 18 December 1940. At the rear is a Henley of No.9 B&Gs.

Right: No.312 Squadron's Hurricane being refuelled by a tractor-towed bowser at Penrhos. The aircraft is Mk I, V7028 DU-V.

Below: RAF Penrhos in February 1941. The nearest Hurricane is that of F/O Adolf Vrana – Hurricane Mk IV6678, DU-L of No.312 Squadron – with DU-S on the right. The Whitleys and Battles belong to No.9 B&GS.

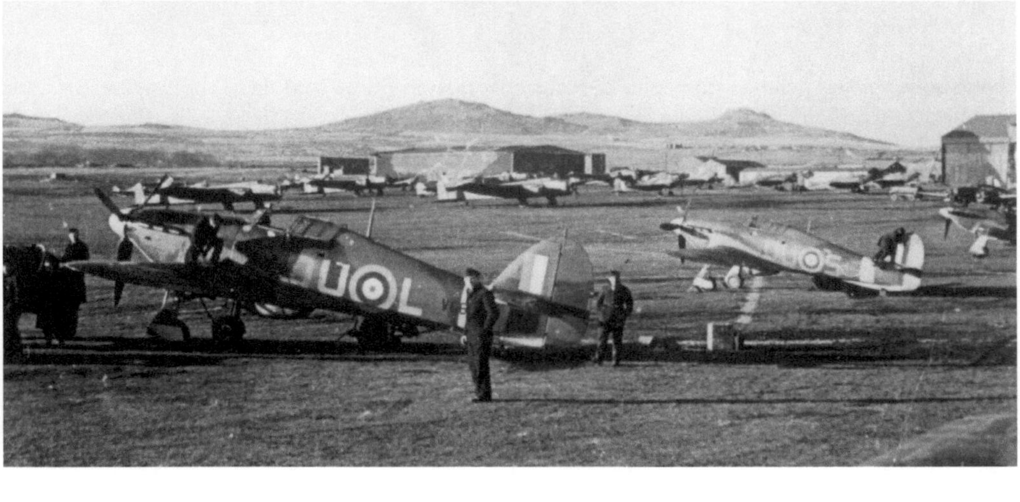

Left: Part of the same view as before but viewed from the cockpit of V6678, DU-L.

Below: Snowed-up in the winter of 1940–41, No.312 Squadron's Hurricane Mk I, V6678 again; Adolf Vrana's personal aircraft.

workshops, the gas defence store, plus the fabric workshop. Some strafing of the marine craft, and Air–Sea Rescue Unit at Pwllheli, also took place.

After a short respite further attacks on the 9th and 10th came. At 6.10 a.m. on the 9th, in low cloud, a JU 88 suddenly appeared and flew over the airfield at a height of 100ft, with guns blazing whilst dropping bombs and incendiaries. Thirty minutes later another aircraft attacked, causing damage to a hangar and the station CO's office. The following day another attempt to bomb Penrhos was made when three aircraft dropped bombs from high altitude. The first salvo fell in marshy ground and did not explode; the second fell at Pensarn near Pwllheli, with no damage other than cratering.

Following these raids urgent protection was needed, so on 12 October 1940 a flight of six Spitfires from No.611 Squadron at Tern Hill was flown in to Penrhos to provide some defensive cover, but as usual it was too late and there were no raids to contend with. Three of the six Spitfires returned to Tern Hill on 21 October. On 21 December the 611 detachment, which had seen no action, was back at its new posting at Southend.

Hurricane Mk I of No.312 Squadron at Penrhos in February 1941. It was grim for the ground crews that year!

Flight of No.1 AACU would move across the bay to Towyn at the end of June 1941. A new RAF airfield had been completed north of Penrhos near Caernarvon. It was to be named RAF Llandwrog and it opened for the use of No.9 Air Gunners School (AGS) on 21 January 1942. All the training of gunnery that was currently carried out at Penrhos was to be moved to this new station and in yet another change of name the unit at Penrhos was to be known as No.9 Observer Advanced Flying Unit from the last day of February 1942.

Before No.9 AGS at Llandwrog could take over the gunnery side of Penrhos's commitment, the policy makers decided that the needs of the observer programme were the most important and Llandwrog was to be a satellite of Penrhos in the work of No.9 (Observer) Advanced Flying Unit ((O)AFU). So No.9 AGS was disbanded at the satellite and all the night flying of the observer courses was flown from the new concrete runways at the satellite. Six Ansons left Penrhos in January 1942 to form a permanent night flying detachment at Llandwrog and gunnery was to be a part of the course.

The view across Penrhos in 2002. The photograph was taken from the hill that has the Battle HQ and air-raid shelters on its slopes. (I. Jones)

Workshops and maintenance building No.8 at Penrhos in 2002. (I. Jones)

The small grass airfield at Penrhos, with its 20ft drop to the surrounding marshy coastal ground, was to see a lot of accidents as the faster and heavier types of aircraft were used. The biplane era had been left behind without any alteration at the airfield at all and many aircraft came to grief here.

The loss of men and machines in the mountains of Snowdonia had increased to an extent that on 29 October 1942 the senior medical officer at Llandwrog wrote, 'during the last five months there has been ten major crashes in the area, which this station sick quarters has attended over an area of 40 x 30 miles. Time taken – one to three days in the mountains. Depending on the degree of accessibility. Total time away from the station fifteen days. Total number of dead forty, total number of injured removed to hospital, eight.'

At 9 p.m. on 14 January 1943 Anson EG110 crashed while on a night flight to Shrewsbury and back. The next day a telephone message was received stating that the pilot had made his way to a farmhouse. The station medical officer and a party set out to search for the aircraft, while Valley and Llandwrog made an air search. The search party gave up at 2 a.m. on the 16th owing to the moon setting and bad conditions on the mountains. At dawn the party set off again and located the wreck at 11.00 a.m., by which time two of the crew were dead and the navigator, who was seriously injured, had been lying in the aircraft for thirty-nine hours.

This particular incident led to a conference on mountain crashes being held at station sick quarters, Llandwrog, when it was decided to allocate a Humber ambulance and a jeep for hill trails with radio equipment. Subsequent crashes brought about more changes and better survival equipment. Eventually the professional Mountain Rescue Service was created on 3 November 1943 at the station of Llandwrog.

The headquarters of No.9 (O)AFU was transferred to Llandwrog from Penrhos at this time because of its runways and modern facilities, making Penrhos the satellite.

The end of the war was in sight and on 25 November No.60 Air Gunnery course, the very last to train here, passed out. The Mountain Rescue Service was transferred to Llanbedr, across the bay, just before the station at Llandwrog closed on 29 June 1945 and put on Care and Maintenance.

No.9 (O)AFU ceased to exist on 16 June 1945 and No.2 Aircrew Holding Unit (ACHU) was formed at Penrhos. The strength of the holding unit was 960, being dealt with by the re-allocation boards of Nos 25 and 29 Groups. On 1 October 1945 No.54 Group assumed responsibility for Penrhos from No.25 Group and on 31 March 1946 No.2 ACHU disbanded. RAF Penrhos passed into history.

At the war's end the airfield carried a lot of pre-war equipment. There was one of the old seaplane sheds, classified as the 'F' shed (drawing No.2723/36). It was the main hangar on site. There were three Bellman hangars (drawing No.8449/37), nine Blister (over) hangars and one extra over Blister. The control tower/watch office was of timber construction (drawing No.1409/36) and appears to be a single-storey hut placed at the northern limit of the airfield on the apron near the 'F' shed. Although the airfield was always grass, there was a 50ft-wide concrete perimeter track surrounding two thirds of the circumference, leaving only the southern limit of the landing area without track. The domestic sites were not dispersed, as they were to be on later airfields, and the living and instruction buildings were all to the north and the east of the technical site.

The runways on grass were north-west–south-east, which was the longest at 2,800ft. The north-east–south-west was 2,300ft and the east–west was 2,500ft.

The domestic sites became a home for Polish servicemen who did not wish to return home after the war. A large caravan touring camp occupies the old technical site and a small airstrip is used by private light aircraft on the airfield proper, mostly in the summer.

There are many air-raid shelters, strong points, ack-ack posts and the Battle HQ on the high ground that lies behind the airfield to the north-east and quite a few of the RAF technical site buildings are extant. The hangar bases are also still to be seen.

The airfield at Hell's Mouth in July 1946. The three Bellman hangars are evident, as is the concrete base that was supposed to support a fourth. The moving target range is between the hangars and the sea. (Welsh Assembly Government Photographic Archive)

FIVE
Hell's Mouth

Grid reference SH 270285, 2 miles west of Abersoch, Gwynedd

The name of this place, as far as the people who live there are concerned, is Porth Neigwl and it is situated on the southern tip of the Lleyn Peninsular. The RAF needed bombing and gunnery ranges for the Armament Practice Station newly built at Penrhos, some 10 miles to the north-east. These ranges were to be later developed into a Relief Landing Ground.

The ranges covered a large area and required the compulsory purchase of seven farms, the bay and the sandy beach that extended 7 miles in length from Cilan Head to Rhiw Mountain. Along this distance targets were erected, usually of 10ft sq. canvas sheets with rings for gunnery training from the air. Offshore were anchored the two target rafts for bombing practice with 8.5lb smoke bombs. When bombing with live bombs took place it had to be done outside the 3-mile limit.

This airfield construction at Penrhos and Hell's Mouth had begun in 1936 and war was not popular or inevitable, bearing in mind that this part of Wales was a hotbed of fervent nationalism, causing the huge protest locally at the imposition of these war-like structures and land purchases.

From its opening in 1937 the range welcomed the aircraft from various Flying Training Schools that were visiting Penrhos for a month's armament practice. In the days leading up to the outbreak of war these were Hawker Harts, Demons and Audaxes, then Gladiators and Battles. The target towers in those early days were the Westland Wallaces and later the Henleys. Air-to-ground, air-to-air gunnery and bombing at high level and low level, plus dive bombing of a shallow variety, took place as a matter of routine when the weather allowed.

The use of the range by early 1940 had been reduced either by the weather, which could be a nuisance when sea mist shrouded the place, or a change in training techniques with the larger, faster aircraft coming into service, but the range was to be altered. The small grass airfield was to be extended, three Bellman hangars and aprons were to be built on the south-east corner of the site and between these and the sea a large moving target railway was to be laid.

By July 1940 the hangars were ready, as was the moving target range, but the extensions to the airfield were not and would not be ready until May 1941.

The moving target range consisted of an oval-shaped course in plan, a light railway with a motorised flat truck that carried a wooden model of an aircraft as it travelled the railway. It was fired at by the trainee gunners, who were installed inside replica gun turrets. The model would be shredded, eventually, by the machine-guns and replaced by a new plywood model when needed. It seems that the idea was successful, so much so that units of the USAAF came over from East Anglia to try it with a view to adopting it. It also brought the anti-aircraft gunners of the airfield defence at Penrhos, here to improve their aim after the string of air raids that Penrhos had suffered.

The personnel at Hell's Mouth were, at this time, one officer, one flight sergeant, two or three sergeants, a few corporals, about thirty airmen and a few RAF Regiment men, all gone by February 1943. Their duties were to operate the bombing range and gunnery targets. The bombers, mostly Ansons, would fly over on a bombing run, aim and drop 12lb smoke bombs which would ignite on hitting the sea. Positioned at points along the bay were the three quadrant towers where observers

Rescue pinnace, built at nearby Pwllheli, that served with the APS at Hell's Mouth. (Via E. Doylerush)

The two buildings are Range Target Control (left) and No.2 Quadrant Tower (right). The erosion by the sea has reduced the space between the sea and the buildings dramatically since the war. In the distance, as marked, is No.3 Quadrant Tower. (I. Jones)

Looking north-west up the rising ground of the airfield; the fence is post-war and there is a considerable area of the airfield behind the photographer. The blockhouse/machine-gun post is to the right of the centre as marked. (I. Jones)

would focus telescopes on the smoke bomb hits and note the compass bearing on a brass ring that was fixed in position in the windows of the quadrants. There were usually three quadrants on duty: No.1 in the south-west corner of the bay, No.2 in the middle, within walking distance of the camp, and No.3 in the far north-west which the two men who manned it had to use cycles to reach.

The Bellman hangars had their hangar 'Queens', apparently. Eyewitnesses among the staff claim that they were reserve aircraft for Penrhos but no-one wanted them. At this time there

Above: Looking north to Towyn Farm. (I. Jones)

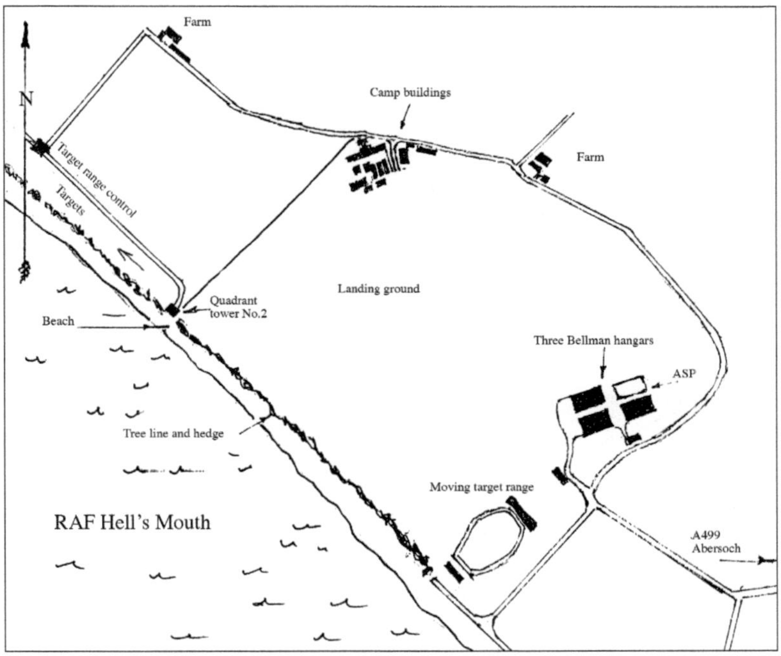

Right: Map of the airfield of Hell's Mouth. (I. Jones)

were four Blenheims and a Whitley in the hangars, being engine-runned and checked over once a month but languishing through the war doing nothing.

During its wartime existence Hell's Mouth was a safe haven for a lot of aircraft in trouble over the Irish Sea or Cardigan Bay, lost in poor visibility or short of fuel. Two of those beautiful aircraft, Whirlwinds, were sent to intercept an enemy bomber in the Irish Sea. They were from No.263 Squadron at Fairwood Common and were forced to give up the chase in heavy rain. Running out of fuel after a long flight, the leader decided to turn east after losing wireless contact with base. Luckily they just passed over the tip of land at Hell's Mouth, saw the field and landed with nothing left in the tanks. This was on 22 May 1942.

Another emergency was the P38 Lightning from Atcham, which was lost; with empty tanks it landed safely on 25 September 1944.

When the airfield's extensions were completed in May 1941 it became a relief landing ground to Penrhos in Flying Training Command.

In June 1942 the new airfield and training station at Llandwrog came into being. Its improved runways and modern facilities altered everything for Penrhos, and more so for Hell's Mouth. They were obsolete.

At the end of the war, where Hell's Mouth was concerned, the RAF wanted to continue using the ranges for bombing and gunnery practice after the war, and indeed were actually thinking of expanding with offshore ranges that would extend 8 miles seaward. This plan caused much local

Trainee gunners on the moving target range. (E. Doylerush)

The concrete road that was laid from the farm track. The bases of the old camp also remain. John Owen, the current farmer, uses the place to dump manure now. (I. Jones)

Blenheim Is were at Penrhos and Hell's Mouth with No.9 Bombing and Gunnery School from 1939 to 1941. There were fifty-three on strength by 1941. Some were pushed into the hangars at Hell's Mouth, never to see daylight again, although the engines were occasionally run up.

An air-raid shelter of Stanton type, near the communal site, near Towyn Farm. (I. Jones)

Right: This is the machine-gun post facing the sea, not far from the camp. (I.Jones)

Below: Westland Wallaces did good service at Hell's Mouth in the early years and were used for towing the sleeves or flags that were used for the targets of air-to-air firing. (I. Jones)

protest that the RAF gave in, although they refused to release any of the land. Caernarfon's MP at the time, Mr Goronwy Roberts, pressed continually for its release and his persistence paid off early in 1947. By April range clearance work had begun to make the land safe for agriculture again.

There is little to see here now, except a few air-raid shelters and a blockhouse. A section of the old perimeter track is used as a minor road. The concrete bases of the Bellman hangers, with their steel floor runners, can be seen, as well as the remains of the moving target range, although this is now derelict. The base of the guardhouse and the entrance road running up to it from the gate can also be seen. The main camp buildings were north of the hangar site, near Towyn Farm and comprised accommodation blocks, messes, latrines and stores of targets, flags and ammunition. All that remains of this site is the road into camp, an air-raid shelter, and a blockhouse.

The present farmer at Towyn is too young to remember, but his parents told him of those wartime days and he is keen to learn more. He and his wife have provided me with copies of the official RAF Target Range maps of Hell's Mouth, along with other material, earning my sincere gratitude.

RAF Llandwrog on 10 May 1946, showing the close proximity of the runways to the sea. This PRU sortie was just six months before the erection of all the hangars on the runways for No.277 Maintenance Unit. (Welsh Assembly Government Photographic Archive)

SIX
RAF Llandwrog

Grid reference SH 435590, 4 miles south-west of Caernarfon, Gwynedd

RAF Llandwrog was intended as a fighter station in a forward position in the advent of an enemy invasion from the Irish mainland, of which there was a fear in around 1940. It was also an airfield that would be useful in the interception of raiders attacking the industrial north-west from bases in north-west France.

The site chosen was on a small peninsular between Caernarfon Bay and Foryd Bay at the western entrance to the Menai Strait, a very historic place with Celtic and Roman sites abounding the area. The airfield is probably the lowest in the UK as it is levelled out at sea level exactly. The main contractor was Sir Robert McAlpine and subcontracts were issued to British Runways Ltd. Local builders also had their share of work.

Work began in September 1940. As completion neared in May 1941 Germany's preoccupation with the USSR had diverted the threat of invasion and lessened the need for a fighter defence base, but there was a crucial need for trained observers, navigators, pilots and training facilities, especially at a training base away from the attentions of the enemy.

Flying Training Command was to take over the new airfield, but even before its official handing over the Oxfords of No.11 Service Flying Training School flew in from Shawbury's muddy grass airfield on 11 June 1941 to use the new concrete runways for their night-flying practice. The first of many accidents at this station occurred when one of the Oxfords of this unit stalled on take-off and dived into the sea, with the loss of the trainee pilot, who became the first airman to be interred in the Central War Graves Commission plot at Llanbeblig cemetery, Caernarfon.

RAF Llandwrog officially opened as No.9 Air Gunnery School (AGS) at the time the school was formed here, on 7 July 1941, in No.25 Group Flying Training Command, equipped with twelve Whitleys and twelve Lysanders. Two days after opening, the one and only visit from the Luftwaffe occurred. A JU 88 came in at 50ft from the south and machine-gunned the parked aircraft, damaging Whitley P5024. The raider made off as fast as it came in; unlike nearby Penrhos, which suffered many attacks, Llandwrog was never to be hit again.

Three of the Whitleys of No.9 AGS were borrowed for operations and a detachment sent to Driffield for the 1,000-bomber raid on Cologne. One of these was posted as missing after the raid.

There were numerous flying accidents here, but the most tragic was the collision between two Whitleys attempting to land at the same time, unaware of each other's presence. One virtually landed on top of the other, wrecking both and causing the death of seventeen young airmen.

The main runway, west–east, was the cause of many accidents and if one of the engines failed after take-off you either fell into the sea, as the western end ran onto the beach, or in the reverse direction, to the east, you could fly into mountains.

As well as Whitleys for turret training and Lysanders for target towing, the unit was using a few Battles and Ansons. When the weather was anything but hopeless they were busy here and the training of gunners was pushed ahead. The real urgency was for observers because of the

Airspeed Oxford R6329 of No.11 Flying Training School from Shawbury, which under-shot at Llandwrog on 10 July 1941. (E. Doylerush)

Whitley of No.9 Air Gunnery School (AGS) at Llandwrog. Some of this unit's aircraft were detached to Driffield for the 1,000-bomber raid on Cologne. One was posted missing after.

long training programme required for this trade. Nearby Penrhos was training observers at No.9 Observers School, but was handicapped by not having hard runways, plus the fact that it was a small airfield. As it was vital to do the section of an observer's course relating to night flying from a decent-sized airfield with hard runways, it was decided to move six of No.9 Air Observers School's (AOS) Ansons to Llandwrog from Penrhos. The unit's night-flying course arrived at No.9 Air Gunnery School (AGS) as a lodger on 21 January 1942.

The trainee gunners attended the Moving Target Range at Hell's Mouth on occasions, to improve their aim by firing from an old turret on the ground at a large, wooden model aircraft, mounted on a motorised truck, running on a small-gauge railway. The usual practice for the gunners was out over the sea along the coast in the turret of a Whitley or Anson, or perhaps a Defiant, firing at a drogue towed by a Lysander. They were, of course, always to the seaward side of the gunners' aircraft. The air gunners' syllabus included much theoretical classroom work. The workings of the Vickers and Browning machine-guns were studied and the stripping and clearing of stoppages practised, the operation of turrets, mounting guns in them, re-arming, the theory of the reflector gun sight and harmonising the guns to them and gravity drop were all considered, as well as aircraft recognition, types of ammunition in use, such as ball, tracer, incendiary, explosive and armour-piercing.

The official takeover of Llandwrog as a satellite of Penrhos came about at the expense of No.9 AGS, which disbanded on 13 June 1942; Penrhos took over some of the gunnery training commitments of the disbanding unit, but only the part that was for observer training.

Lysander IIs of No.9 AGS. These were used as target-towers at Llandwrog. (Frank Jones)

The western end of the 1,100-yard main runway (No.08), just 20 yards from the sea. The mountains to the east are Llyn Dulyn and the surrounding minor hills of Llandwrog. (E. Doylerush)

On 1 May 1942 No.9 AOS was redesignated No.9 (Observer) Advanced Flying Unit ((O) AFU) at Penrhos. Flying Training Command had not only the home-based pupils to train but men who had been trained abroad and were to be taught the complexities of navigating and flying, day or night, summer or winter, in the UK. The Empire Air Training Scheme was adopted and observers on the first course of No.9 (O)AFU took part in the first course of the Air Training Scheme.

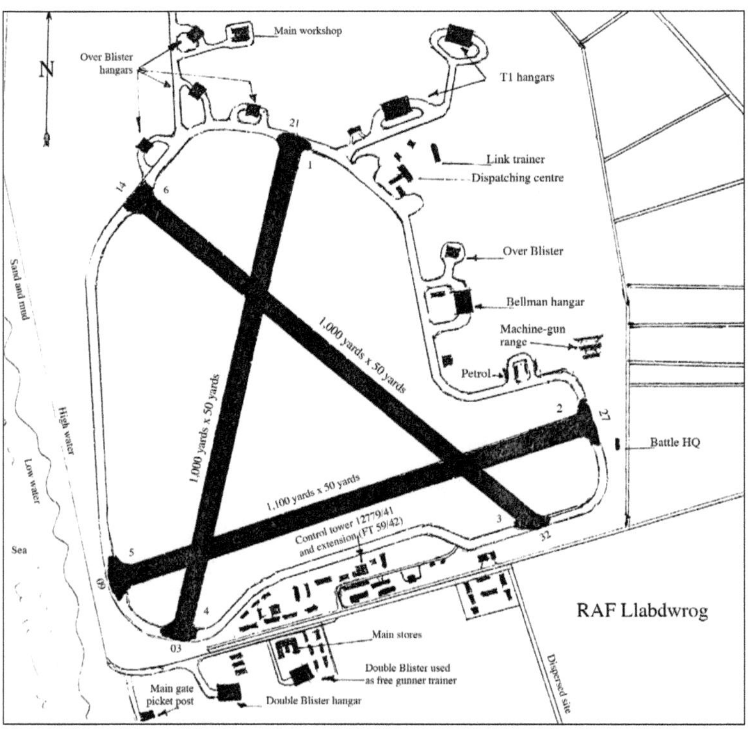

Top: The buildings that survived until 1978, the when it was decided to make the civil airport and museum at Llandwrog. (E. Doylerush)

Centre: Even the turrets of the Defiants were used by the trainee gunners of No.9 AGS for gunnery practice at Llandwrog.

Left: Map of RAF Llandwrog illustrating the runway lengths and buildings. (I. Jones)

RAF control tower at Llandwrog (drawing No.12799/41) prior to its complete refurbishment for civil use.

The Whitleys were all withdrawn and replaced by the Blenheim Mk IV in late 1942 for duty with No.9 (O)AFU at Llandwrog.

The many accidents at sea from downed aircraft resulted in Llandwrog having its own small Navy. A jetty at Fort Belan, a few miles away, on the tip of the peninsular, was the harbour for several pinnaces engaged in air-sea rescue for the whole area.

By mid-1942 the number of personnel at Llandwrog exceeded that of Penrhos and, as the facilities at Llandwrog were more modern and extensive, the roles of the two stations were reversed; Llandwrog received the headquarters of No.9 (O)AFU with Penrhos now the satellite. The personnel at Llandwrog at this time comprised a pupil intake of eighty-six, pupil output of

Above: Avro Anson of No.9 (O)AFU at Llandwrog. They suffered many losses – thirty between 1942 and 1943. Causes were the weather, high ground, poor navigation and pilot error.

Left: Llandwrog in the period September 1946 to December 1955, when No.277 MU occupied the airfield. Note how the runways are packed with T2 hangars. (Welsh Assembly Government Photographic Archive)

forty-four, those still training 122; station strength was thirty-five officers and 577 airmen. No.9 AOS detachment had eight officers and 136 airmen. Following this change the Whitleys were withdrawn from service and replaced by Blenheim Mk IVs. Ten flew in on 3 August, followed by another eight on the 4th. Ansons were used for navigation training and there were many casualties with these. From 1942 to 1943 thirty aircraft were lost due to a combination of factors, such as being caught in bad weather, flying into high ground, inadequate navigational aids and pilot error. Training went on at Llandwrog without much change through to the war's end.

The rapid expansion of flying training in the area had also increased the number of high ground crashes in Snowdonia. Penrhos was the station that was notified in the early years when an aircraft was lost in the mountains because it was nearest, but Llandwrog was to be the station that would take charge of this brave and exhausting work. At Llandwrog the equipment was developed and perfected to save many lives in harsh, unforgiving conditions. The Mountain Rescue School was founded here and set an example to many other schools.

No.9 (O)AFU churned out observers until disbandment on 14 June 1945 and on 29 June RAF Llandwrog closed for flying. Personnel at war's end numbered 1,313: 134 officers plus four WAAF; 298 NCOs plus five WAAF; 713 Airmen plus 159 WAAF.

The refurbished control tower at Llandwrog, now known as Caernarfon Civil Airport. RAF building No.108 (drawing No.12779/41).

Aero Club aircraft alongside the control tower. This is the current scene at Llandwrog. (I. Jones)

The Mountain Rescue School was later transferred across the bay to RAF Llanbedr, relinquishing accommodation blocks into which No.2 Aircrew Holding Unit moved. This unit was to process many Canadians for release before its job was done.

A most extraordinary task was to be given to this little-known station. From September 1946 until late 1955 RAF Llandwrog was to be host to a maintenance unit, No.277 MU. In those years it was to receive 9,000 tons of dangerous enemy chemical weapons. Under the name 'Operation Sandcastle', these weapons were lorried from Germany to a channel port, then shipped to the docks at Newport, Monmouthshire, and loaded back onto lorries to be driven all the way to the

The new hangar that was built for civil aviation. The museum is to the left of picture. (I. Jones)

The RAF watch office at Llandwrog. (I. Jones)

Lleyn Peninsular, to this remote airfield. There, after processing of some kind, they were again loaded onto lorries and driven to Fort Belan, where a jetty had been built. They were transferred to Tank Landing Craft (TLC) and sailed to a point near Stranraer where a ship, the *Empire Clair*, was waiting to be loaded. This hazardous cargo was then deposited into the deepest part of the Atlantic.

The processing of these chemical weapons was dangerous and everyone wore masks inside the base. Children living locally were instructed to keep hold of their wartime-issue gas masks long after people elsewhere had handed theirs in. The Ministry of Defence clearly had a commitment to getting rid of these weapons. However, with all the loading, and unloading, the distance from Newport to Llandwrog, which in those days was a long way to travel, and the fact that these weapons tended to leak begged the question – why here?

To enable the work to be done safely, a lot of hangars were erected on the runways here. Initially, eight were placed on the north-west–south-east runway, then eventually all the runways had become occupied by a total of twenty-two Bellman hangars. There were a lot of young RAF men, mostly doing their post-war national service, and a number of civilians working on this large enterprise. When the gas and liquid toxic material had been extracted, it was loaded aboard lorries again for the trip to Fort Belan.

One wonders now whether everyone knew what was really going on at this airfield, especially in view of the Welsh Nationalist protest at the building of these military airfields in 1938.

The Airfield

The usual three runways: the main, west–east, 27/09, 1,100 x 50 yards; north-west–south-east, 21/03, 1,000 x 50 yards; north-east–south-west, 14/32, 1,000 yards. All were of concrete and tarmac.

At the end of the war, before Operation Sandcastle, there were two T1, one Bellman, five Over Blister and one double Blister hangars. The air traffic control tower (drawing No.12779/ 41) with an extension (drawing No.FT59/42) sat to the south of the main runway and about halfway along its length. The watch office was alongside (drawing No.18441/40).

Now known as Caernarfon Airport, the management is by Snowdon Mountain Aviation and it provides flying instruction and pleasure flights.

All the hangars have been demolished but a new small hangar was erected by the company Keenair of Liverpool, just after the Royal Investiture of Charles to Prince of Wales, when Caernarfon was quite busy with the media. On weekends it is still busy with private light aircraft.

RAF Bodorgan in May 1946. The satellite landing ground is to the south and the gap in the tree-line can be seen. Aircraft to be stored here would fly to the airfield proper – to the north – and then be tractor-towed to the SLG across the road. (Welsh Assembly Government Photographic Archive)

SEVEN
RAF Bodorgan

Grid reference SH 385685, Anglesey

This station opened under the name of Aberffraw on 11 September 1940, a simple grass airfield that was intended to provide a base for the aircraft that were to perform the Anti- Aircraft Co-operation work in this area of many artillery batteries and schools. This field was later to evolve into a satellite landing ground.

Built on the coast near Maltreath Sands, between the villages of Hermon and Aberffraw, in the remote south-west of Anglesey, the land was requisitioned in December 1939 from Sir George Meyrick of Bodorgan Hall. Sir George protested long and hard, but was forced to accept the situation, as had many other landowners in the UK. The local working people were pleased at the thought of the work that the new airfield would bring them.

The first unit to arrive at the new station was 'Z' Flight of No.1 Anti-Aircraft Co-operation Unit (AACU) on the 11 September 1940 from Watchet in Somerset, followed in January 1941 by 'J' Flight of the same unit from Penrhos. Both of the flights were to operate the 'Queen Bee', the clever, remote-controlled target version of the de Havilland Tiger Moth which was meant to be flown pilotless over the gun batteries and be fired at with live ammunition, to give practice to the gunners, mainly those of Ty Croes School of Artillery.

For some reason the Queen Bees of 'J' Flight were handed over at this time and 'J' Flight was re-equipped with Henleys, Tiger Moths and Martinets; perhaps it was felt that one flight of these unpredictable aircraft was enough on one station.

Another unit at Aberffraw was No.2 Maintenance Flight, which became No.3506 AAC Servicing Unit in December 1943. It serviced the Henleys at this station and those of other flights from nearby airfields such as Morfa Towyn and Penrhos. All three of the above units were part of No.1 AACU and Aberffraw was to be in No.70 Group from December 1941.

The main task for the Queen Bees here was to satisfy the target needs of the Ty Croes Range. It was important to Anti-Aircraft Command that all ack-ack schools turn out competent gun crews to meet the raids that the enemy was increasingly inflicting on the country, so places like Ty Croes and Manorbier were to be given top priority.

The first pilotless flight from here took place on 2 December 1940. Queen Bee P4804 was airborne for two and a half hours before crashing on landing, which was always the most dangerous part of the exercise for the controller on the ground. His skill as an operator was very important. He would bring the Bee down to 15ft from the ground, a trailing line connected to the magneto would touch the ground causing earthing and thus stop the engine. The Bee would then land of its own accord. The operator's control panel was fairly simple – a face with three buttons marked left, straight and right, and diagonally across the face were four more marked climb, level, glide and dive. Radio equipment in the Bee received control signals from the ground and transferred them into movements of the rudder, elevators and ailerons. This radio set was prone to partial or complete failure, particularly after a near miss from the artillery when the aircraft would become uncontrollable and either crash or head out to sea on a course built into the system upon failure.

Westland Lysander II of No.6 AACU. This unit carried out night flying from Bodorgan in an attempt to aid the anti-aircraft gunnery night shooting.

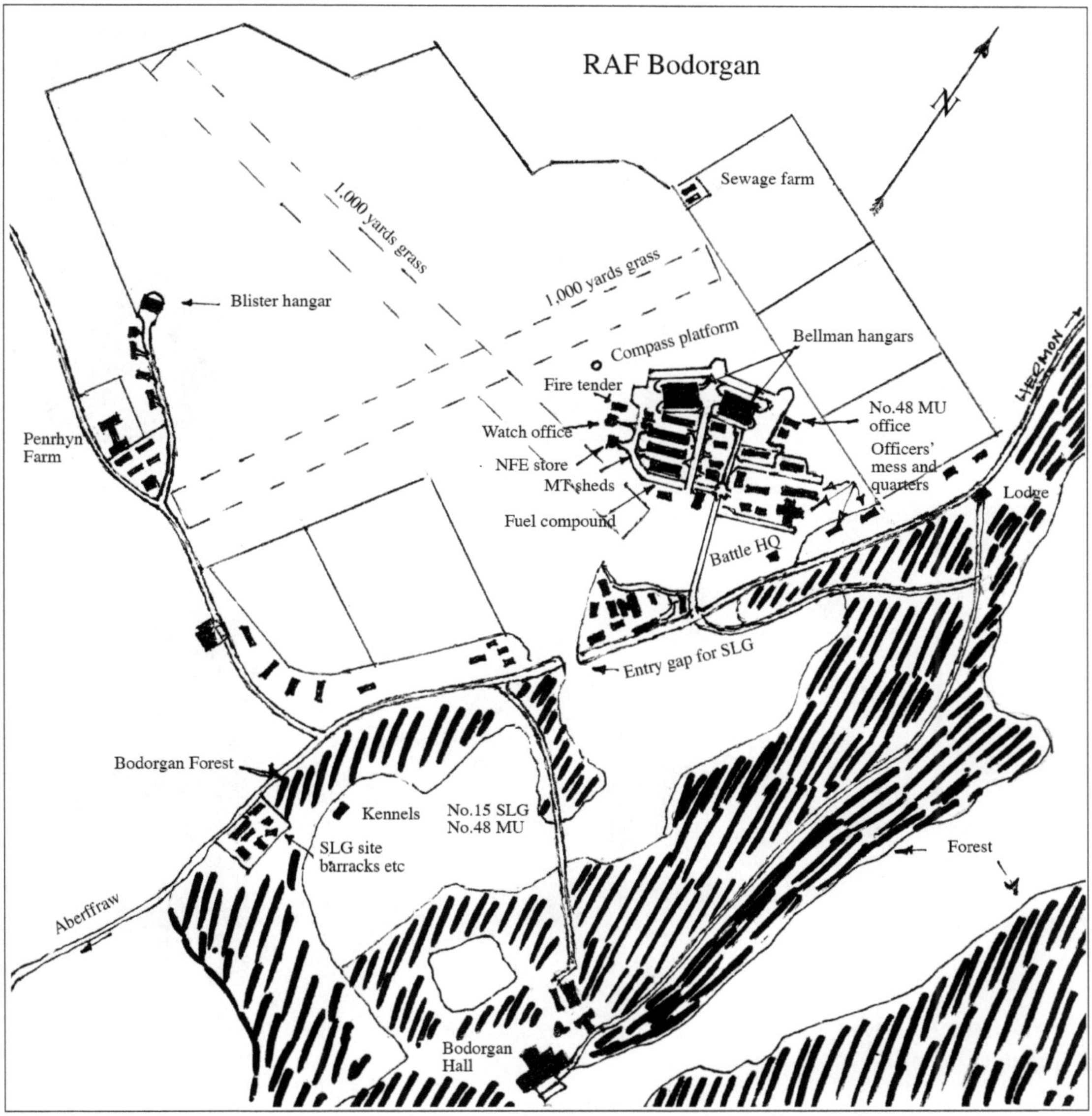

Map showing the airfield buildings and runways (grass) and also the SLG in the forest to the south. (I. Jones)

Opposite centre: Hawker III of 'J' Flight at No.2 Maintenance Flight for servicing at Bodorgan. This unit was heavily engaged with the repair and maintenance for all Henleys based in North Wales and the north-west of England.

Opposite below: DH Queen Bee radio-controlled target aircraft. These pilotless biplanes were the first aircraft to arrive at Bodorgan when the station opened in September 1940. They were with 'Z' Flight of No.1 Anti-Aircraft Co-operation Unit (AACU), together with 'J' Flight of the same unit.

There are few buildings left at Bodorgan, so it behoves us to photograph what is left. This is the motor transport shed and yard on the technical site, building No.14 on the site plan. (I. Jones)

The DH Dragonfly of No.6 AACU arrived at Bodorgan in July 1941 to fly night exercises from here, again to aid the anti-aircraft artillery. They also flew from Valley.

The AMWD (Air Ministry Works Department) store at Bodorgan, building No.38 on the communal site. (I. Jones)

Soon after operations with the Bees began at this airfield it was found that side winds made things difficult when coming in from the sea and after P4801 crashed from a height of 300ft in February 1941 an instruction was issued: 'In view of the peculiarities of the local weather, pilotless flights will be restricted to those days on which conditions are not approaching the borderline for a safe approach.'

Another problem, which arose during the summer of 1941, was a series of losses caused by battery failure. Following the loss of P5738 on 20 August, when it sank off Ty Croes, every battery used in a Queen Bee had to undergo a vibration test and one hour's piloted flight while still being controllable from the ground. However, V4788 crashed at Glantraeth, Anglesey, in September, again from battery failure. After this every aircraft had to be fitted with two batteries in parallel.

A detachment of Lysanders from No.13 Squadron at Hooton Park practised their night flying from here during the full-moon period of March 1941. Further Lysanders and many other types were to arrive here when this airfield received aircraft for storage on becoming No.15 Satellite Landing Ground (SLG) for Hawarden's No.48 MU.

On 15 May 1941 the name of the station was changed from Aberffraw to that of Bodorgan, in order to rectify delays and confusion, such as the misposting of airmen and the receipt of goods not ordered. The SLG was, of course, using the airfield for its landings and take-offs, but the dispersal of the aircraft was to the south-east in the area around Bodorgan Hall. A large gap was cut in the hedges and trees at the perimeter to allow access to the woods for concealment. Wellingtons were beginning to arrive at this time but the number of aircraft stored was not to exceed thirty at any one time.

Any Queen Bees hit by gunfire and then not responding to the ground controls were designed to fly out to sea until they crashed. This usually worked, but on 24 February 1942 V4793 was hit by artillery fire and refused to answer to the controls from the ground. The operating crew thought that the aircraft had flown seawards and forgot about it. In fact, it was damaged to the extent that the automatic control did not function and it flew on in a southerly direction until it crashed into the slopes of Snowdon.

July 1941 saw the arrival of a de Havilland Dragonfly and a Westland Lysander from No.6 AACU to carry out night flying with gunnery exercises from here and Valley. Officers of the USAAC visited in August to study the operation of the Queen Bees.

On 1 October 1942 'J' and 'Z' Flights became Nos 1606 and 1620 Flights in a reorganisation that affected all Army co-op flights. No.1606 operated Henleys, Tiger Moths and Magisters. No.1620 continued with the Queen Bee.

The boredom of target-towing sometimes led to the pilots doing silly but exciting stunts, usually beating up the gun positions on the ground. On one of these occasions, in September 1942, Henley L3301 collided with a 3.7in AA gun at Ty Croes. The aircraft crashed into the sea and no trace of the crew was found until five months later, when some of the pilot's clothing was washed up on a nearby beach.

A similar accident occurred on 11 February 1944, involving a Martinet at Holyhead. Martinet HP183 was providing co-operation for Bofors guns at 1,000ft with 2,000ft of tow. The co-operation was successful, but the aircraft flew over the gun park, losing height. The aircraft's starboard wing struck a gun and the aircraft turned upside down, crashed and slid for 300 yards. The pilot was killed and the drogue operator was slightly injured.

Three days later another Martinet was engaged on Army Co-operation 3 miles west of Valley, when a Mosquito approaching from the sea collided with the drogue towing cable, despite a red flare being fired by the Martinet pilot. The Mosquito lost 7ft from its starboard wing in the collision but landed at Valley successfully.

December saw yet more reorganisation of anti-aircraft resources in No.70 Group; No.1620 Flight left Bodorgan, whilst No.1606 Flight remained, with eight target-towing aircraft and one Tiger Moth.

Army Co-operation work was carried out at other schools and batteries, as well as Ty Croes. At Llandudno was the Coastal Artillery Practice Camp with Batteries Nos 553 and

Ablutions Block, building No.36 (drawing 16468140) on the communal site at Bodorgan. (I. Jones)

Hurricane IIcs were used by No.577 Squadron for gun-laying and radar exercises for the ack-ack school at Ty Croes. They operated from Bodorgan from January 1944 to January 1945. (This is an example aircraft only.)

No.577 Squadron adopted the Vultee Vengeance IV as a target-tower to give the ack-ack school a faster target than was possible with the Henley. (This is an example aircraft only.)

534, there was No.44 Battery and the 5th Cheshires, plus the Defence Course at Bodorgan to serve.

The Defence Course was one organised by No.70 Group to give RAF officers and NCOs some knowledge of command in the field Army style, in case of emergencies. The first course was in July 1943 and the last in May 1945.

During wet weather, target towing was difficult from this grass airfield and the shooting programme was often delayed. In October 1944, for instance, conditions were so bad after prolonged wet weather that the airfield was closed and operations were moved to Valley. Three Martinets were to go there and a Bellman hangar, office and store made available for use. The unit brought its own fuel to Valley and transport was laid on for crews to and from there each day.

By the end of 1944 the dispersal of aircraft from Hawarden was no longer necessary and No.48 MU's requirements were more than met by the takeover of Hooton Park as a Subsidiary Storage Site. By 30 December 1944 there were no aircraft in store and No.15 SLG was shut down. The arrival of another unit took up the space left vacant at Bodorgan. This unit was No.650 Squadron, an Army Co-operation squadron which had formed at Cark on 1 December 1943 by merging No.1614 Target Towing Flight with 'D' Flight of No.289 Squadron, equipped with Martinet 1s, Hurricane 1Vs and Tiger Moths, and was coded T7. They were to stay and disband here on 26 June 1945.

At this time a detachment of No.577 Squadron arrived, again in the Army Co-operation role. It had formed at Castle Bromwich on 1 December 1943 and flew Hurricane IIcs, Spitfire Vbs and Vengeance IVs, coded 3Y.

No.577 Squadron's detachment left in the new year of 1945 and all the commitments remaining to Bodorgan were carried out by No.650 Squadron This was to slowly run down as Anti-Aircraft Command's needs lessened, until No.650 Squadron disbanded on 26 June 1945 and No.3506 Servicing Flight left for Weston Zoyland in Somerset. In July No.70 Group disbanded and Bodorgan came under the control of No.12 Group. The airfield was closed to all traffic from 23 August 1945 and went on a Care and Maintenance basis in October. The buildings were used to house Italian POWs who had been put to work on nearby farms and the Air Training Corps attempted to acquire the airfield proper as a base for No.63 Glider School, but the redoubtable Sir George Meyrick was having none of it and wanted his land back.

The hangars have long gone, but many other buildings survive and are in use for industry and office accommodation. There were two Bellman hangars on the technical site and one Blister hangar isolated on the south-western perimeter. No.48 MU's No.15 SLG occupied land to the south and east, around the woods of Bodorgan Hall. From what I can find there was no control tower, as such, but a bungalow-type watch office (drawing No.935/40).

The average numbers of personnel mid-war were: twenty-six RAF and two WAAF officers; twenty-three RAF and four WAAF NCOs; 479 RAF and 124 other ranks.

The airfield's grass runways were: QDM 023, 1,000 yards; QDM 080, 1,000 yards; QDM 068, 960 yards.

There are still a few buildings to be seen on the technical site and domestic site at the main gate. The grass airfield seems strangely unchanged nearly sixty years after commissioning.

RAF Valley a year after opening. The sand around the runways is still disturbed and loose on this date, 15 April 1942. Trewan Sands GCI Station is believed to be the buildings at the top right. (Welsh Assembly Government Photographic Archive)

EIGHT
RAF Valley

Grid reference SH 305755, Anglesey

In an area of sand dunes and marshy lakes between the villages of Valley and Rhosneigr in north-west Anglesey the Air Ministry decided to build the badly needed Fighter Command Sector Station for the brand new No.9 Group.

The airfield was needed to defend the exposed cities of Liverpool, Manchester and the Mersey. Since the enemy occupation of France, the Luftwaffe would fly over St George's Channel and the Irish Sea to hit the shipping and towns of the north-west.

Construction of RAF Rhosneigr, as it was then named, was begun by levelling the sand dunes along the coast. This, however, scraped away the roots of the Marram grass that had stabilised the sand and was to have long-term and expensive repercussions in the near future. The civil contractors building the airfield had trouble with the wind blowing sand, as it would drift over their work and into engines.

The opening up party arrived on 1 February 1941. Three runways had been laid by this time: north-west–south-east, 1,600 yards long x 50 yards wide; north–south, 1,300 yards x 50 yards; north-east–south-west, 1,400 yards x 50 yards. Very little else was ready, however.

To this uninviting place came the first fighters – 'A' Flight of No.312 Squadron's Hurricanes from Speke on 3 March 1941. They were to stay only until 18 April 1941 when they moved on to Jurby and were replaced by the Hurricanes of No.615 Squadron, which were supposed to be on a rest.

On 7 May one of No.615 Squadron's pilots, Flt Lt Haywood, who was on patrol over the Menai Straits, saw an enemy aircraft at 10,000–10,500ft above him. He immediately attacked, but three seconds after opening fire his guns jammed, possibly with sand. Nonetheless, hits were made before the enemy escaped by diving into cloud.

A week later Sgts Hamilton and Roberts contacted and engaged a JU 88 while on convoy patrol, 10 miles west of Holyhead. Hamilton silenced the rear gunner and after running out of ammunition, broke off the attack, by which time his aircraft had been damaged by return fire and he himself had three bullets in his leg. Roberts's aircraft was also damaged and his port guns were out of action. The consolation was that the enemy aircraft was badly damaged and both the Hurricanes landed safely.

On 3 April the station was renamed Valley and the name Rhosneigr was deleted from the files.

The next combat took place on 27 June when Flt Lt Haywood and PO Strickland intercepted a JU 88 20 miles off Holyhead. A few bursts of fire caused the raider to dive out of sight. Bullets were seen to strike the starboard engine so the aircraft was claimed as damaged.

Things were quiet until 26 August when a JU 88 flying over Cardigan Bay was attacked by FO Mouchotte and Sgt Hamilton of 'B' Flight, resulting in damage that forced the Junkers to land in Eire, near Cork. Also in pursuit was 'A' Flight who had investigated a 'Friendly' aircraft and were too late. This was the last action for No.615 Squadron in North Wales. No.615 departed

The sole Gaydon hangar at Valley, on the main technical site. (I. Jones)

These are the Hurricanes of No.312 (Czech) Squadron, which were based at several Welsh airfields in turn. The squadron served at Valley for just six weeks, with its 'A' Flight detached here in April 1941.

in September 1941 and was replaced by more Hurricanes, in the hands of No.242 Squadron, but were made non-operational in October to prepare for an overseas posting. They left Valley in December.

These squadrons had provided daytime cover for the region, but night cover was also needed. In May 'A' Flight of No.219 Squadron was detached from Tangmere with four Beaufighters to provide this cover.

They claimed one success on 1 June when Sqn Ldr Colbeck-Welsh, on patrol over Anglesey, shot and crippled a JU 88, which crashed into the sea near Aberystwyth. At the end of the month the flight returned to Tangmere and was replaced by a detachment from No.68 Squadron, based at High Ercall. They used Valley as an advanced base for their night-fighting Beaufighters.

By this time, mid-1941, five radar stations were built around the sector at strategic points – Castell Mawr, Rhuddlan, South Stack, Nefyn and Pen-y-Bryn near Aberdaron. Radar coverage

No.615 Squadron was at Valley with its Hurricanes from May to September 1941, before leaving for another Welsh station, Angle. This photograph shows the squadron earlier at Manston.

Bristol Beaufighter of No.68 Squadron. This was a night-fighter unit that was based at High Ercall but used Valley as an advanced base.

was supplemented by the GCI (Ground Control Interception) station named Trewan Sands which was close to the airfield and came under its control.

The night fighter squadron that stayed the longest at Valley was the one formed here on 30 June 1941, the Royal Australian Air Force No.456 Squadron. Equipped initially with eighteen Defiants, they learned the art of GCI and became operational on 5 September 1941. It took this long because of a shortage of aircrew and the majority of these being British rather than Australian.

There were so many training units and flying schools in the area that flying accidents were increasing. This was apart from operational casualties, so No.275 Squadron was formed at Valley in October 1941 for air–sea rescue duties, with Lysanders, Walruses, Defiants and Ansons.

No.350 (Belgian) Squadron also formed here. This was redesignated the Belgian Flight of No.131 Squadron, to be a Belgian National Squadron equipped with Spitfire IIas, Vas, Vbs and Vcs, and was financed by the Belgian Congo Fighter Fund.

Spitfire V of No.350 Squadron. This was a Belgian squadron that had formed here at Valley on 12 November 1941.

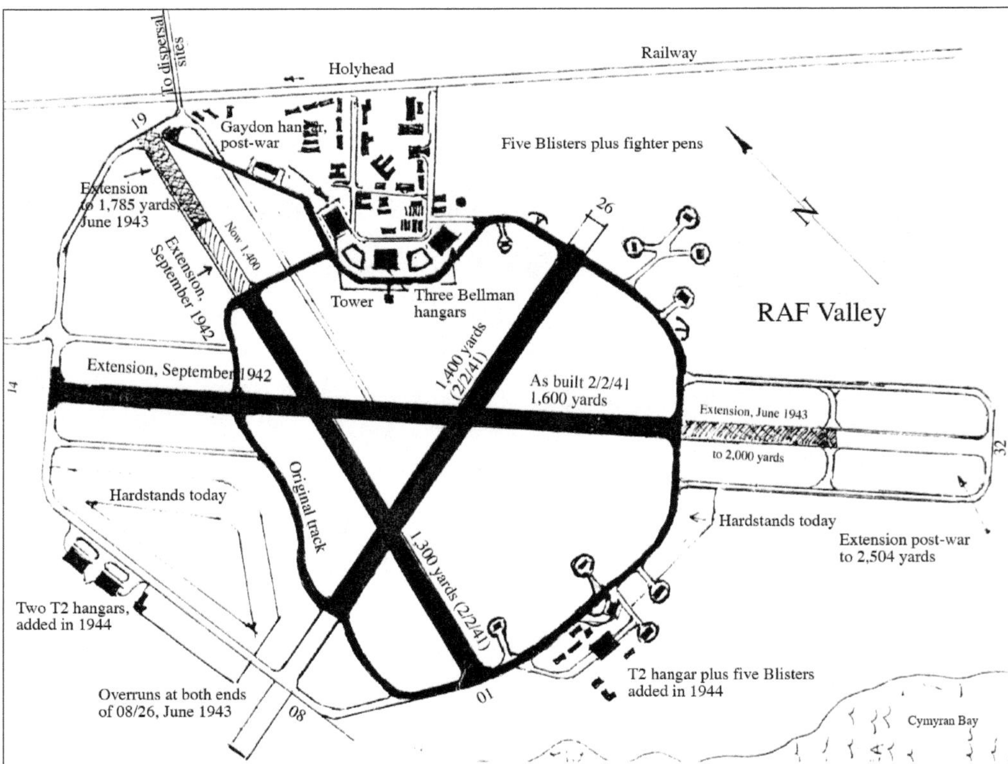

Map indicating the runway lengths and extensions, hangars and other buildings at Valley. (I. Jones)

Beaufighter IIF T3370 RX-Z of No.456 (RAAF) Squadron at Valley in the summer of 1942.

No.68 Squadron's Beaufighters had had a quiet spell until 12 October when PO Mansfield and his radar operator Sgt Janocek had a busy time with no less than five contacts, all JU 88s. Under the guidance of Trewan Sands GCI the first contact was at 10 p.m. whilst flying north over the Irish Sea, but the enemy escaped.

After further vectors from Trewan he made contact with the second JU 88 thirty minutes later. At a range of 300ft the Beaufighter opened fire with two short bursts. Flames appeared on the aircraft's port wing and it dived steeply, disappearing into cloud. PO Mansfield claimed this as a 'damaged'.

The next combat was at 11 p.m. when Mansfield opened fire at a range of 250ft but he did not observe any hits. He then had to take avoiding action to prevent a collision, following which the JU 88 made his escape.

At 11.25 p.m. the fourth contact of the night was made at a height of 12,000ft. Closing in gradually to firing range, PO Mansfield kept the enemy in view silhouetted against the moon before giving two one and a half-second bursts which set the starboard wing and engine alight. It then disappeared from view into cloud. Coast Guard and ROC observers reported a fire burning on the sea at the time of the attack, so the JU 88 was claimed as destroyed. Five minutes later another JU 88 was shot down by Mansfield in almost the same manner. It fell, burning furiously, into the sea. All these interceptions took place in the area of the Skerries.

The squadron's next success was on the 22nd when FO Winward, on patrol at 4,000ft, was vectored by Trewan Sands to intercept an enemy aircraft, which turned out to be another JU 88. He successfully attacked this aircraft from a range of 250ft, resulting in a brilliant white flash – probably exploding flares – from the fuselage. Winward watched the enemy aircraft plunge into the sea near Nevin.

No.68 Squadron had more success on 1 November 1941 when at 9.30 p.m. PO Shipherd and his radar operator Sgt Oxby were on patrol over Bardsey Island at 10,000ft. Trewan gave vectors for an enemy at 7,000ft, Shiphard made radar contact and closed in on his quarry. He closed slowly to 1,000ft and saw a silhouette that identified it as a Heinkel 111. Closing to 400ft and easing to dead astern, 20ft below the German, he opened fire with a two-second burst. The enemy caught fire and the whole interior was ablaze. The plane spiralled down into cloud, then exploded when the plane hit the ground at Bwlch-y-Fen Bentir Farm near Bodffordd.

The airfield was still suffering from the effects of sand, which was getting into everything. Workers' lunch boxes, engines and lorries were bogged down if left for long. Large quantities of steel matting were laid, but attempts to stabilise the sand failed because the seed was blown away with the sand.

The Spitfire Vbs of No.131 Squadron at Valley had trouble with sand getting into their Merlins and so returned to Llanbedr.

On 25 November No.68 Squadron left Valley to return to High Ercall, leaving the night defence of the sector to No.456 RAAF Squadron, which was now fully operational. The squadron had dropped the Defiants by October and re-equipped with Beaufighters, but not without some difficulty as the gunners on Defiants had to be re-trained as observer's radio so that they could use AI equipment installed in the Beaufighter.

In the autumn of 1941 Valley was allocated a decoy airfield at Newborough Warren. It was a 'Q' site and was about 7 miles to the south-east of Valley. Four men and a corporal ran this site from an enlarged Anderson Shelter. Their work entailed the switching and testing of lights, clearing the lamps of sand and the maintenance of generators.

The Belgian No.350 Squadron was formed here on 12 November 1941 by redesignating the Belgian Flight of No.131 Squadron. This, made up of many experienced pilots, a lot of whom had escaped from Europe after flying against the enemy, did not take long to become operational. Its duties were to be mostly convoy patrols and were not sorry to leave Valley for Atcham on 2 April 1942.

Other units based at Valley during 1941 for short periods were two sections of Hurricanes of No.302 Squadron and one flight of Spitfires from No.403 Squadron in June and July respectively, and the Hurricanes of No.605 and No.43 Squadrons in September.

No.456 RAAF Squadron's night fighters had their first victory on 10 January 1942, when Sqn Ldr J. Hamilton and POD Norris-Smith were scrambled to intercept six enemy aircraft suspected of laying mines in the sea approaches to Liverpool. The time of scramble was 10 p.m. and Hamilton kept losing contact with the enemy due to poor communications with Trewan Sands GCI, arising from faulty RT equipment. Eventually GCI Hack Green took over the interception control. A few minutes before midnight Hamilton made contact with a Dornier DO217 at 12,000ft and after a seven-minute chase, shot it down with two bursts of cannon fire.

Another squadron that spent a brief time with Valley was No.131 Squadron with Spitfire Vs. They had moved up from nearby RAF Llanbedr early in March 1942, in order to operate patrols in the Irish Sea. As soon as the squadron arrived it began to suffer the effects of wind-blown sand, finding its way into the engines and guns with remorseless efficiency. This caused endless serviceability problems. Indeed, all squadrons operating from Valley had this serviceability trouble until the problem was solved.

No.131 was operational on 12 March when Flt Lt Harris and Sgt Vilboux (Free French Air Force) took off in their Spitfires to intercept a JU 88 30 miles south-west of Valley. They successfully attacked the raider and sent it into the sea with both engines on fire; there were no survivors.

By mid-April the sand had defeated No.131 Squadron. The problem had become so acute, with engines having to be overhauled far more frequently than normally, that arrangements were made to return to Llanbedr immediately.

A No. 247 Squadron Hurricane IIc. This unit was at Valley from September 1942 to February 1943.

The new Station CO, Wing Commander W. Churchill DSO DFC, had to solve the problem of the sand and after a discussion with interested parties, a suggestion to dredge silt from nearby lakes and spread it over the airfield was voiced. This simple remedy was put into effect and, to the relief of all, it was found to work very well. It stabilised the sand, allowing the grass to grow, which, in turn, further stabilised the sand.

No.315 (Polish) Squadron arrived at Valley in May 1942 and on 23 August engaged a JU 88 near Dublin, killing the gunner and inflicting damage that caused the aircraft to force-land near Wexford. This combat ended as a draw, as a Spitfire was also shot down.

The night fighters of No.456 Squadron had some action on 30 July 1942 when Wing Commander Wolfe and P/O Ashcroft took off from Valley shortly after midnight to patrol Cardigan Bay. Whilst over the Lleyn Peninsular they were given vectors for two enemy aircraft, but as there were OTU aircraft on exercises in the general area Wolfe was cautious. Having made contact with the nearest of the two aircraft, he carefully identified it as a JU 88 before launching an attack. However, the enemy pilot had seen the Beaufighter and, after violent manoeuvring, managed to shake off his pursuer. Wolfe now turned his attention to the other raider still in the vicinity. After obtaining a visual sighting at 2,000ft range, he identified it as a Heinkel He 111. Closing to 250 yards he fired two short bursts causing the enemy to hastily jettison his bombs. More gunfire resulted in flames coming from the bomber. Wolfe broke off the attack and watched the stricken aircraft go into a dive and fall 2,000ft to crash on Pwllheli beach. Two men had bailed out and were later picked up from the sea (coded F8 + lw). This was the last German aircraft shot down over North Wales.

The favourable weather record at Valley had attracted many diversions, both RAF and USAAF. It was chosen to be one of the airfields to receive the American four-engined bombers that were anticipated in the UK. The station was to stay in No.9 Group Fighter Command but Ferry Command would use it as a lodger unit.

In September another detachment of Hurricanes arrived. This was from No.247 Squadron from High Ercall, taking over from No.315 Squadron until February 1943.

During September and October plans were drawn and approval granted for a large expansion of the airfield – a 360-yard extension to the south end of the north-west–south-east runway, bringing it to a length of 1,960 yards, and an extension of the north–south runway from 1,300 to 1,400 yards at the north end. In addition, the ground at both ends of the runways was to be levelled to make overshooting and undershooting less hazardous. Also, in 1942, a small air-to-ground firing range was prepared at Carmel Head on the north-east coast of Anglesey, for the use of visiting squadrons in conjunction with No.1486 Fighter Gunnery Flight. This flight was formed by redesignating No.9 Group Target-Towing Flight and was equipped with two Martinets, two Masters and three Lysanders. After its spell at Valley

This overhead shows the improvement to the airfield at Valley, now that the sand is under control. Note the runway extensions of September 1943. (Welsh Assembly Government Photographic Archive)

it was to move to Llanbedr, where it was disbanded on 18 October 1943 to become No.12 Armament Practice Camp.

No.275 Air–Sea Rescue Squadron was busy all the time. It was out almost every day, retrieving aircrew from the sea. The unit had Spitfires and Ansons to join the Walruses and Lysanders now. The Ansons were used for dinghy and supply dropping. In June, for example, apart from exercises and training flights, No.275 Squadron was actively searching for ditched aircraft on ten occasions. The squadron suffered its own casualties sometimes, too, as on 7 May 1943 when PO Furlong, while practising sea landings in his Walrus 2 miles north of Anglesey, got into difficulties because of heavy seas and cross winds. The aircraft lost a float, capsized and then sank. An Anson also sank in July after crashing into the sea, luckily without casualties. No.275 Air–Sea Rescue Squadron moved to Warmwell, Dorset, in April 1944.

No.456 Squadron RAAF were having a lean time and were bored with no enemy activity, faced with routine patrols and exercises with calibration runs for GCI stations. During the summer of 1942 they were re-equipped with the more powerful Beaufighter Mk VI; however, they were not to use them for very long, as they were to convert to Mosquitoes. In January 1943 the conversion took place smoothly and at the end of March No.456 Squadron transferred to Middle Wallop after twenty-one months at Valley.

To replace them came No.406 (RCAF) Squadron from Middle Wallop on 31 March 1943, with Beaufighter Mk VI fs. There was little activity operationally and so the squadron was

Mosquito NF11 of No.456 (RAAF) Squadron spent twenty-one months at Valley. It had formed here on 30 June 1941 and flew Defiants, Beaufighters and Mosquitos, before departing on 31 March 1943.

employed on practice attacks on troops and vehicles near Newborough and Aberffraw. Another task was to impart their experience and operational technique to another night fighter unit – the USAAF 414th Night Fighter Squadron from Honiley. Equipped with the Beaufighter Mk VI fs, they stayed at Valley for eleven days.

No.406 RCAF Squadron left for Exeter in November 1943 after eight months at Valley, being replaced by No.125 Squadron in mid-November with Beaufighters. In early 1944 the squadron was to re-equip with Mosquitoes prior to departure to Hurn in March. The increase of inward flights from the USA was making itself felt by all at Valley and, before their leaving for Hurn, No.125 Squadron was often called to shepherd in the airfield's lost B17s and B24s.

The USAAF Ferry Terminal had been made operational in June 1943 and the role of Valley as a day- and night-fighter station declined with the absence of enemy incursions, so its new role as a ferry terminus from the US flourished. Once the facilities had been updated the Americans loved it and it would appear that it had become their first choice. Even before Prestwick, Air Transport Command of the USAAF was determined to have the station enlarged, especially the domestic accommodation, as the trickle of the US aircraft that were diverted from Prestwick because of the more favourable weather at Valley increased, sometimes at very short notice. Valley was often totally swamped; American servicemen ended up sleeping in hangars and there were no extra food rations to give them a meal when they arrived.

A second transatlantic terminal was needed. Squires Gate had been used, but was not at all what the Americans wanted. They wanted an airfield that satisfied six particular requirements: (1) located on the west coast; (2) with good road and rail links; (3) a favourable weather record; (4) located near to the US Depots at Warton and Burtonwood; (5) possessing good runways, capable of extension if necessary; (6) in an area free from obstructions. Valley met all six.

The RAF also used Valley as a ferry terminal and a diversion airfield because of its good local visibility, when all around was clamped.

The incoming four-engined types from the US would arrive in squadrons or groups, stay overnight or a little longer, then fly to their airfields of operations after flight plans had been plotted to mostly East Anglian destinations.

Valley was still under No.9 Group Fighter Command and some of the senior officers in the RAF were concerned that they were going to lose Valley to Transport Command. Long arguments ensued, but it was obvious that the majority of the activity at the station was in transport and RAF units were thin on the ground.

When No.125 Squadron left Valley in March 1944, it was replaced by No.157 Squadron from Predannack. As it turned out, this was the last fighter squadron to serve here. They carried out defensive patrols over the Irish Sea, but the main objective was to re-equip and train on Mosquitoes fitted with Mk X AI. By early May 1944 they became operational and left for Swannington in Norfolk with their Mosquito NF11s.

The question of Valley's transfer to Transport Command came up again in 1944, by which time its use as a transatlantic terminal had grown enormously. Headquarters of Air Defence of Great Britain (which Fighter Command had become in November 1943) had to concede that the interests of the station were now predominantly transport and, furthermore, the American presence was the major factor in its continued existence.

A large building programme was then launched, further lengthening runways, laying new taxi-ways and fifty hardstandings, and an American Radio Range for Homing and instrument let down was installed. Ferry Command and the US Transport Command shared joint facilities and worked in harmony for more than two years.

An RAF commitment to retain a toe-hold in Valley in a small way arrived on 1 November in the form of No.1528 BAT (Beam Approach Training) Flight. It was formed here and its purpose was to train pilots in the use of blind approach systems. The unit had six Oxfords, two Link

One of the T2 hangars on the west site, with improved door gantries. The two T2s face each other at each end of the aircraft servicing platform. (I. Jones)

The second T2 on the west site at Valley, again with modern gantry. (I. Jones)

The tremendous movement of US Air Force aircraft, in and out of Valley, was impressive in 1945 and '46. Here a B24 Liberator lands at Valley to prepare for the Atlantic crossing in its turn.

Trainers, ten flying instructors and an intake of thirteen pupils every ten days. This unit was to be re-named No.1528 Radio Aids Training Flight on 15 September 1945.

More and more American units arrived at Valley. They built their own Mess Halls and sleeping quarters. Detachment 'F' of the 24th Region Army Airways Communications System arrived on 23 June and also the entire motor pool on the same day. Two officers and eight men arrived to set up the weather station and a PX was operating by 2 July.

The RAF gave up two of the three Bellman hangars to the Americans and from 1 November 1943 Valley ceased to be a sector station. All personnel involved were transferred to RAF Woodvale, which took over control of the sector. Trewan Sands GCI station remained open, but Newborough Warren 'Q' site was abandoned.

The total number of American aircraft handled at Valley in 1943 was 365 – 226 four-engined heavies and 139 twin-engined aircraft.

In 1944 the flow increased. In August 1944 alone 353 four-engined and 144 twin-engined aircraft arrived. As many as ninety-nine B17s and B24s landed on 16 September! The monthly figures for transiting crews during September was 1,588 officers and 2,322 airmen.

By February 1945 signs of movement in the opposite direction were evident. War-weary aircraft and crews were being returned to the States, yet in March, with the European war in its final stages, there were still 518 arriving for the month. On 20 April the last sizeable inward movement took place when sixty aircraft landed and the final curtain fell when eight 10th Air Force Liberators landed.

Since July 1943 Valley had handled over 5,000 US aircraft: 2,734 B17s, 1,869 B24s, 271 C-47s and 174 other types, making an exact total of 5,048.

Ten days later project 'White' was initiated. This was the pre-planned movement of all US military aircraft that were to return to the US after action in Europe was concluded. It was arranged that each Bomb Group would, in turn of length of service in this theatre, be funnelled though Valley and dispatched with its crews to a base in the US and then on to a predesignated airfield, where preparations were to be made for service in the coming offensive against Japan.

Fairey Fireflash guided missiles being loaded onto Supermarine Swift F7s of No.1 Guided Missile Development Squadron (GMDS) at Valley on 16 October 1958.

Fireflash missiless loaded aboard, the Swift F7s take to the sky for exercises with GMDS at Valley in April 1957.

By the end of August 1945 this had been done and some 2,500 aircraft had been through Valley on their way home.

With the Americans gone, No.1528 BAT Flight carried on and was joined by a detachment of Sea Otters of No.281 Squadron that had arrived earlier in March but left for Northern Ireland in October. The BAT Flight received an increase of establishment to eight Oxfords and was re-named The Radio Aids Training Flight (RAT), before leaving in December for Blakehill Farm. On 29 July 1946 Valley was transferred to Flying Training Command.

No.10 AGS reformed here on 1 December 1946. No aircraft were allocated; it was an organisation to warrant Valley's Diversionary Airfield status and was to stay until 1951.

No.63 Glider School came to stay in January 1947 from Tal-y- cafn, but was gone by April 1948.

Although things looked bleak for Valley at this time, there was no doubt amongst the upper echelons of the RAF that, with all the expenditure by the USA on this station, it had to have the permanence that it deserved. It was put on a Care and Maintenance consistent with its diversionary status, and then No.2 Pilots' Refresher Flying Unit was formed here on 6 August 1947 from an element of No.21 (Pilots') Advanced Flying Unit, with Oxfords and Harvard 11bs. This was disbanded at Valley on 1 April 1948, to be absorbed by No.1 RFU.

No.20 Squadron arrived on 19 July 1949 in the Anti-aircraft Co-operation role from RAF Llanbedr, equipped with Spitfire 16s, Beaufighter TT10s, Vampire F1s and F3s, continuing their work with the Artillery Schools around the West Wales coast, until disbanding here on 16 October 1951.

March 1951 saw the formation at Valley of No.202 Advanced Flying School with Vampires. This unit was redesignated No.7 Flying Training School on 1 June 1954 and was equipped with Vampire FB5s and FB9s, Anson 19s and Chipmunk T10s. A year later this school was joined by No.6 Joint Services Trials Unit, a lodger unit that was established to undertake trials with the Fairey Fireflash air-to-air missile, that was redesignated No.1 Guided Weapons Squadron in 1957.

This unit was unique in flying the only fleet of Supermarine Swift Mk F7s modified to carry the weapon. In January 1959 the unit was renamed yet again, becoming No.1 Guided Weapons Training Squadron.

On 5 August 1960 No.4 Flying Training School was formed at Valley by redesignating No.7 FTS. It was named No.4 (Advanced) FTS from December 1966, with Vampire T11s, Varsity T1s,

Vampire FB5 VZ 874 of No.7 Flying Training School at Valley in October 1956.

Folland Gnats, two-seat trainers of No.4 FTS at Valley on 29 April 1963.

Folland Gnat XR949 of No.4 FTS taking off at Valley, 1965.

Hawker Hunter F6 XF384 of No.4 FTS at Valley.

Hawk TI of No.4 FTS at Valley flying over the Brittania Bridge at Bangor.

No.19 Squadron's Hawks on dispersals at the west site at Valley, 2002. (I. Jones)

Gnat T1s, Jet Provost T4s, Hunter F6s and T7s, and eventually Hawk T1/1As. The Central Flying School also had a detachment flying Hawks at Valley, tasked with training pilots to be instructors for the RAF Fleet Air Arm and other services.

No.1 Guided Weapons Training Squadron was disbanded in May 1962, to be replaced by the Fighter Command Missile Practice Camp (later Strike Command Missile Practice Camp), still in residence at Valley.

In 1962 a unit that was to be a lodger here arrived. This was the Central Flying School helicopter detachment, which was later replaced by No.2 SAR Squadron of No.2 FTS. This unit became the Search and Rescue Training Unit in 1979, before moving to St Mawgan.

Squadron office and HQ for No.19 Squadron at Valley.

The new air traffic control at Valley (drawing No.5871/55). (I. Jones)

Robb portable hangar on the west site at Valley. (I. Jones)

Now at Valley are the Flying Sea Kings of 'C' Flight of No.22 Squadron. No.4 Flying Training School remains in residence, with two component squadrons – No.74 R and No.19 R – and RAF Mona is the Relief Landing Ground.

The Airfield

The runways, as originally laid out in 1941, have been described, but the joint discussions with the USAAF, the last being in June 1943, brought about a huge expansion at Valley. Practically the whole of the station was reorganised in one way or another and runways were extended. An extension of 40 yards was made to the south end of the 14/32 north-west–south-east runway, making its total length 2,000 yards, and a 1,000-yard extension was made to the north end of the north–south runway, making it 1,785 yards long. The third runway was left untouched.

Fifty hardstands were added and improved taxi-ways built. The construction of two T2 hangars on the western site were completed, together with a large apron. This is how the airfield was laid out at the war's end.

When the station was transferred to Flying Training Command and the anticipated fast jet aircraft that were to operate from this base were arriving, the runways had further work done on them. Another extension to the southern end of the main runway, 14/32, was to make it a total of 7,500ft long. A new Gaydon hangar was erected on the site of one of the old Bellman hangars. A new control tower (drawing No.5871c/55) was built a short distance from the wartime original, which, of course, is disused now.

RAF Mona on 13 August 1945, just two months after disbandment and closure. At this time it was under a Care and Maintenance order. This was to continue until it was adopted as a relief landing ground for Valley in 1953. (Welsh Assembly Government Photographic Archive)

NINE
RAF Mona

Grid reference OS 415760, 3 miles west of Llangefni, Anglesey, on the A5

This station was first known as Heneglwys, but after consideration of the problems caused by the Welsh naming of earlier stations, it was decided to name this one Mona, the Roman name for the area. It was also the site of a First World War Royal Naval Airship Station that was named Llangefni. Those stationed here performed the duties of anti-submarine coastal patrols from this bleak spot on the highest part of this central point of Anglesey up until 1918. Thereafter the buildings on the site were used as an isolation hospital known as 'The Druid', after a nearby farm. By the time of negotiations to requisition the area of land that encompassed the hospital the old huts were badly deteriorated by the damp and windy weather on this high and exposed place.

The land was requisitioned in 1941 and the building of the airfield went on through 1942. In its agreement with local government officials and the hospital committee, the Air Ministry was obliged to bear the costs of replacing the hospital at nearby Llangefni. RAF Mona was opened in December 1942 under No.25 Group (Armament) of Flying Training Command, with a station commander of group captain rank.

The original intention was to form a new air gunners' school (AGS) here, to be named No.6 AGS, with a proposed establishment of thirty-four Ansons and twenty-seven Martinets, catering for a trainee population of 240. Such a unit was formed here on 22 October 1942, but no aircraft were allocated. This unit was disbanded on 6 November, just two weeks later, in favour of bringing an existing gunnery school to Mona. No.3 Gunnery School moved in from Castle Kennedy.

A collection of aircraft arrived at Mona, including Blackburn Bothas that had been relegated from operations to training. Between 8 and 24 December 1942 forty-eight of these unpopular aircraft were ferried in from Castle Kennedy, plus six Battles, eight Martinets and two Ansons. The staff pilots of No.3 AGS included many Poles who had escaped to Britain.

Training commenced on 3 January 1943 with forty pupils from the No.2 Signals School, who made up the No.1 Course of Wireless Operator/Air Gunner. At this point in the war training courses lasted for seven weeks, with an average of sixty pupils per course. Air firing drill was carried out from the turrets of the Bothas, and later from the Ansons, firing at drogues towed by Martinets. On the ground another part of training was clay-pigeon shooting. Strangely, this was considered to be very good for learning the basics of quick response and allowance for leading the target. It was also cheaper than aerial firing practice.

In February No.5 (P) Advanced Flying Unit (Special) Flight came to Mona from Tern Hill. This flight, equipped with Masters, comprised twelve Turkish officer pupils, a liaison officer, plus maintenance and instructional staff. The task of this small detachment was to teach the current RAF gunnery techniques to the Turks. They returned to Tern Hill on 16 March. Later in September and October some Czech and French pupils were taught. In 1944 a number of Norwegians and Belgians also had training here.

Two of the station's Bothas crashed during February 1943 – W 5029 in the process of landing on the 15th and W 5024 on the 28th. Luckily there were no casualties. A Martinet of No.3 AGS

Top: Arrester gantry and capstan, situated at the western end of the main runway at Mona. This was installed in the post-war period to save aircraft and pilots when trouble came their way. (I. Jones)

Centre: The south-west end of the main runway at Mona, the only runway in a good condition. Note the arrester cable on the ground. (I. Jones)

Left: This is a two-print mosaic photograph of RAF Mona, under construction, on 15 April 1942. A dotted line indicates where the runways would be and how the road at the bottom of the photo was to be moved down to that line. (Welsh Assembly Government Photographic Archive)

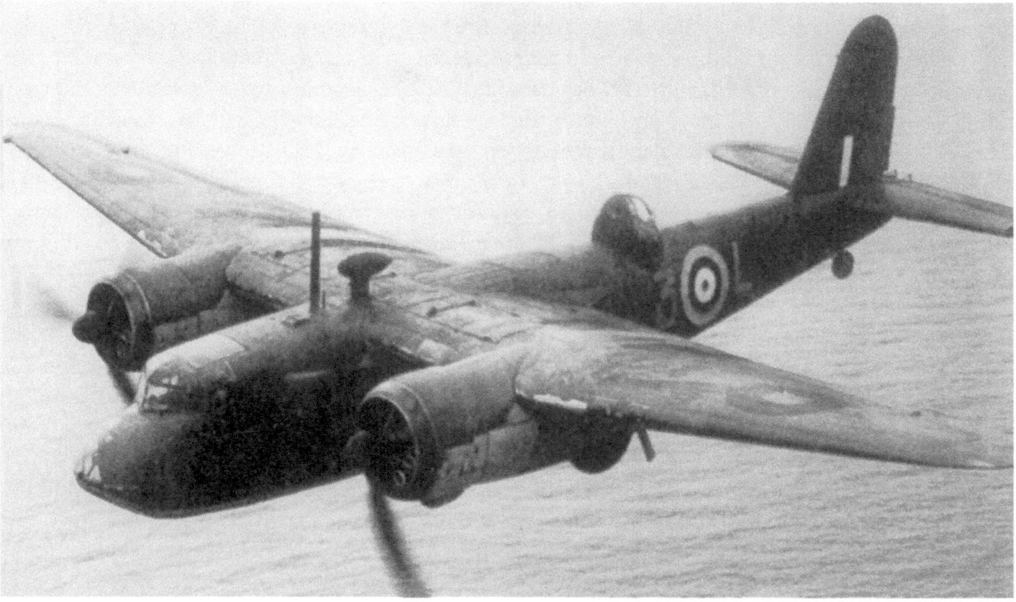

Botha L6318 of No.3 AGS, Mona, in August 1942.

Miles Master II of No.5 (P) AFU (Special) Flight at Mona in February 1943.

Miles Martinet of No.3 AGS. These were the target-towers for the turret gunners and the air-to-air firing from fighter planes.

suffered a mishap when its engine cut following mismanagement of the fuel supply by the pilot, who made a wheels up landing in a field neat Aber. Two other Martinets suffered similar fates at this time. The first was Martinet HP623 from No.61 OTU which force-landed on Prestatyn Sands on 3 April after incorrect handling of the fuel levers. Then, on the 4th, Martinet HP267 of No.1486 Gunnery Flight from Valley force-landed on Cymyran Beach adjoining the airfield. All these cases were due to the pilot failing to change tanks correctly.

In May thirty-one airwomen arrived to undergo a new training schedule. Balloon barrage airwomen were re-mustering as trainee flight mechanics and the preliminary course for this was at Mona.

From May 1943 the Bothas of No.3 AGS were gradually replaced by Ansons and training continued through the summer. On No.26 Course sixty pupils were passed and then, in October, news arrived that the unit was to return to Castle Kennedy at the end of the month. On the 29th ten Ansons and ten Martinets left, to be followed in November by the remaining aircraft – thirteen Ansons, fifteen Martinets, one Master and one Tutor. The Station Headquarters' staff at Mona remained as the nucleus of the next unit to arrive.

This unit formed at Mona – No.8 (Observer) Advanced Flying Unit – was the last (O)AFU to be formed during the Second World War and the opening took place on 15 November 1943. The purpose of the unit was to extend the usefulness of observers by training them in other aircrew duties, such as wireless operation, navigation, air gunnery and bombing. Ansons were the only aircraft used; forty-one at first, to train eighty pupils. Later, this was increased to 200 trainees and an extra four Ansons, bringing the total to forty-five. Two of these Ansons were damaged beyond repair on 25 January 1944 when a severe gale struck the camp, resulting in the collapse of two Blister hangars in which the aircraft were housed.

Avro Anson I of No.3 AGS at Mona. Note the replacement Bristol turret that had ousted the Armstrong Whitworth one, in order to give the gunner a quicker response.

Anson I of No.8 (O)AFU at Mona between November 1943 and June 1945.

Above: Private aircraft parked in front of the last remaining T1 hangar of the original three. (I. Jones)

Right: Air traffic control tower at Mona (drawing No.12779/41). (I. Jones)

Below: Between the hangar and the tower lies the pupil's locker and crew room. This was the original (drawing No.6a1851/41). (I. Jones)

Bomb aiming was carried out at Lligwy Bay on the north-east coast of Anglesey, where two observation quadrants were used to plot hits and misses, operated by ten airmen and a sergeant, all billeted in local houses. Another practice target was Conway Bridge, to which an infrared device was attached. This system allowed practice bombing of pinpoint targets, even in an urban area, and worked by projecting an infrared beam along a previously decided line and angle necessary to score a hit. Aircraft bombsights had linked photographic plates, sensitive to the infrared end of the spectrum and, assuming the attacking aircraft was on the correct course, when the bomb release button was pressed an image was recorded on these plates. The plates would then be developed at Mona and examined to see whether a hit had been achieved on the bridge or how near the miss was. Navigator pupils learnt Morse radio procedures and the techniques of navigation by dead reckoning. The main training area was North Wales and the Irish Sea, with coast-hugging trips down to South Wales also a prominent part of training. After the successful completion of their courses the pupils would go to OTUs.

All that remains of the north-west–south-east, 14/32, runway at Mona. (I. Jones)

All that remains of the north–south, 18/36, runway at Mona. (I. Jones)

As Mona was only 6 miles away from Valley, where the transatlantic terminal for the bombers and transports of the US Army Air Force was based, it sometimes happened that the pilots were confused as to which of these two airfields was their correct destination and made landings at Mona, until traffic control officers at Mona watched their approach on radar and gave them a timely signal to 'keep moving'.

In January 1944 target-towing Martinets from nearby RAF Bodorgan made Mona their home, as muddy conditions at their own airfield made landings, and any movement at all, on the grass field difficult and their Army co-operation work was falling behind schedule. They were later forced to use Valley as well.

Another target-towing commitment began in April 1944, when a detachment of No.776 Squadron (Royal Navy) arrived at Mona. Consisting of two Blackburn Skuas and a handful of naval personnel, they were tasked with providing target-towing for the gunners of naval motor torpedo boats based at Holyhead.

Training continued uninterrupted at Mona until the end of the war, with the same unit – No.8 (O)AFU – still in place. Flying ended officially on 19 May 1945. Disbandment came on 14 June and Mona went on a Care and Maintenance basis.

After a few years of lying dormant the airfield came into use again as the relief landing ground for Valley in the early 1950s. It was not necessary to maintain three runways, so the two shorter ones were closed and the main runway was extended to 6,000ft, with the addition at the northern end. This was to handle the Vampires of No.202 Advanced Flying School that were based at Valley, then the Gnats of No.4 Flying Training School, then the Hawks of today.

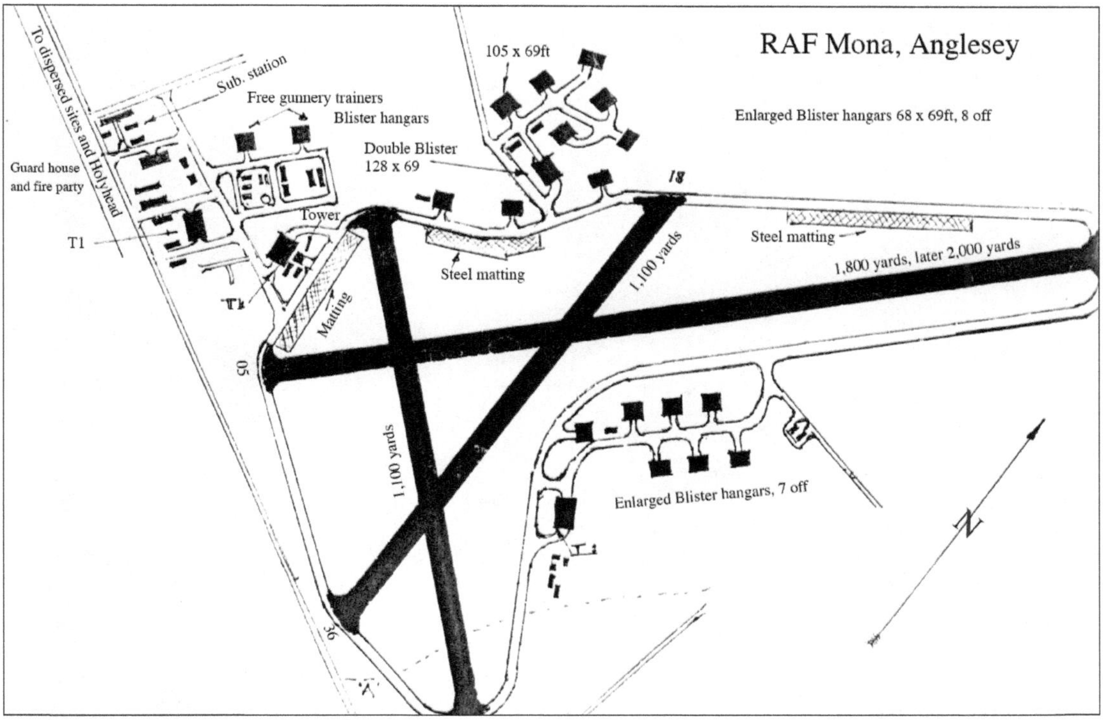

Map of the airfield and buildings at RAF Mona. (I. Jones)

There is a private 'Mona Flying Club' based here now, using the only hangar left of the many that were here during the war.

The Airfield

The airfield's three runways, as built, were: north–south, 18/36, 1,100 yards; north-west–south-east, 14/ 32, 1,100 yards; north-east–south-west, 05/ 23, 1,800 yards, later extended to 2,000 yards.

There were three T1 hangars and seventeen Blister hangars, 68ft x 69ft, and two enlarged, one of 120ft x 185ft and another of 105ft. x 69ft. It seems the building of the many, and larger than usual, Blister hangars was an emergency move when it was discovered that one could not leave aircraft exposed outside due to the salt-laden damp atmosphere that prevailed, high up on this sea-facing island. Two of the T1s were on the technical site to the east and the third was on the perimeter track to the south-east.

An unusual feature was the steel-matted hardstands at three locations around the perimeter track.

There were ten dispersed sites, including the technical and airfield. Most of the accommodation sites were distributed along both sides of the A5 towards Holyhead.

At the end of the Second World War the personnel was: 110 (RAF) and eight (WAAF) officers; 220 (RAF) and ten (WAAF) NCOs; 1,048 (RAF) and 390 (WAAF) other ranks. This gave a total of 1,378 RAF staff and 408 WAAF.

On a visit in July 2002 there was little left of the buildings of wartime. The flying control tower (drawing No.12779/41) is not only still there but still in use. The solitary T1 hangar, the main runway with its arrester gear intact and quite large portions of the other runways are still evident, but with industrial development in its stride on the south-east of the airfield all these parts of the disused runways will be gone soon.

The Saunders-Roe facility at The Fryars, Beaumaris, in August 1945. (Welsh Assembly Government Photographic Archive)

The T2 hangar showing the large extensions on its west side. In the background to the left is the B2 hangar and the boiler house on the far left. (I. Jones)

TEN
Saunders-Roe at Beaumaris

Grid reference SH 610775, Anglesey

Although this establishment was a civilian venture and its work was mostly confined to seaplanes and flying boats, Beaumaris was to be an airbase. Although there was no airfield, there was the water of the Menai Straits.

When the Saunders-Roe Co. was awarded huge contracts at the beginning of the Second World War, as were other companies in the UK who were aviation manufacturers, Saunders-Roe needed to disperse a lot of their work to safer places than their factories at East Cowes on the Isle of Wight, which was well within the sights of the German air force. One of their contracts was to be the conversion of the American Consolidated Catalina flying boats to British specification, for their use by anti-submarine squadrons of the Coastal Command.

A search for a suitable site that would provide a good area of quiet water for take-off and alighting, then a safe mooring for large numbers of these large aircraft was led by the company's test pilot and others. Most of the places that would have served had been snapped up by 1940 and were in the hands of other manufacturers or fulfilling the needs of the Navy, Air Force and Army. It was, apparently, the Ministry of Aircraft Production (MAP) who suggested that Saunders-Roe (Saro) look at the Menai Straits.

The site was perfect – not only deep water and safe anchorage but, furthermore, the Straits ran from the south-west to the north-east with the prevailing wind. To cap their relief, on the Anglesey side of the Straits, a mile to the north-east of the historic town of Beaumaris, was a country house and estate called 'Fryars' which was up for sale and would therefore not need the wretched process of compulsory purchase. Moreover, there were no military airfields nearby, thus giving Saro freedom of local airspace.

Saro bought the estate for £ 5,250 in December 1940, but the Design and Drawing Office, plus the Research and Development departments, had been there since the autumn. The house was large and was used for these offices and white collar departments. The need for hangars and workshops was critical, so one of the Saunders-Roe hangars was dismantled at Cowes and reassembled at Fryars. The MAP then came up with a T2 hangar, which is still on the site.

The land at Fryars runs down a significant slope to the road (B5109) that runs along side the sea edge and added to the later problems that occurred when attempting to manoeuvre the Sunderlands and Catalinas from the sea to their place of work. Another snag was the lack of buildings of any kind that could be used for accommodation for the workers or for storage of parts and materials. Even the pier at the resort at Beaumaris was used as a parts store. The first Catalinas, two in number, arrived in April 1941, but because the slipway was not complete and no hangar or technical staff were available, the first conversions were done afloat. This situation never was to be repeated and, on completion of the slipway, Catalina AM266 was hauled ashore on 28 April.

Saunders-Roe was to operate under the auspices of the Directorate of Production, Canadian and American. This body was to supervise all aircraft imports from North America and to ensure

This oblique gives a reasonable view of the hangars at Saro's Beaumaris. Taken from the south-west, it has the house of Fryars on the extreme right and many workshops, improvisation being the rule at this critical part of the war.

The Catalina is the type one would usually see at this spot in the gardens of Fryars, awaiting a place in the hangars or needing a small modification only.

that modifications needed to bring these imports to RAF specifications were carried out, and Saro was to be responsible for the seaplane and flying boat side of this need.

Some of the early work was tackled at Greenock, on the Clyde, by Scottish Aviation, but once the hangars were erected at Fryars this work was stopped and all work on the Catalina, Mariner and Coronado flying boats from the USA was to be done in Anglesey. The company already had a commitment to supply wooden hulls for the Supermarine Walrus Mk II ASR amphibian. Saro was so short of space that they used local garages, garden sheds; the stables of the Bryn Hyfryd manor house became the Embodiment Loan Store. Space was not only utilised at Beaumaris but also Menai Bridge and Bangor.

Saunders-Roe had built a wide slipway opposite the house at Fryars with a view to doing some of the work on the Short Shetland flying boat, so when the first of the Catalina IIs arrived from the States, the slipway was ready for them. Skilled workers were also being directed here as more workshops became available. At this time, the company was introduced to the Vought-Sikorsky Kingfisher and the Seamew floatplane.

A Catalina being hauled out of the water, up the slipway, at Beaumaris.

The Vought-Sikorsky Kingfishers were prepared to Navy specifications at Beaumaris.

By October 1942 the company had total responsibility for the introduction into RAF service, through conversion kits and then flight testing, of the Catalina, Coronado, Mariner, Seamew and Kingfisher. The conversion kits were manufactured in the workshops at Mona and were not only fitted to the aircraft here but delivered to other bases abroad and to MUs.

The Coastal Command Development Unit had a permanent presence at Fryars and a pool of test pilots from the services. Their work included Leigh Light fitting (where the wing needed strengthening), ASV radar, Magnetic Anomaly Detector (MAD), RAF radio and bomb racks to take a heavier bomb load fitting, and many alterations to the hull's interior for stowage of items that experience with war in the Atlantic had brought.

The Mariner and Coronado did not meet Coastal Command's requirements and after considerable modifying and testing, the programme was halted. During 1944 twenty-two

Martin Mariner JX1 17, after damaging a float in the Menai Strait, lies awaiting repair at Saro's Beaumaris.

Another Mariner, with radar above the cabin, lies high and dry on the lawn. (Via D.J. Smith)

The Catalinas would park along the Strait from the Menai Bridge to Fryars. Forming a queue, the nearest to Fryars was the next one to go up the slipway, and so on.

The Catalina Mk 3 amphibian was eventually rejected by the RAF for conversion as the undercarriage space needed for the retraction of the wheels took up too much room inside the hull. (David Price)

Mariners out of the twenty-six delivered to the UK were ferried back to the United States, but the Catalinas were so good that political efforts were made to get more of them over here quicker, as the delivery was spasmodic. Soon the Catalinas were to be seen at their moorings all the way to the Menai Bridge. Ten were converted in 1941, not nearly fast enough for Coastal Command, and seventy-seven were completed by April 1942.

When the Catalina was brought up from its place in the queue, way down by the Menai Bridge, it was brought up the slipway, across the road, passed through the gap in the wall that surrounds Fryars, and then hosed down with fresh water. All loose equipment was removed and conversion could begin. If space was not available ashore for some time and the aircraft were at their moorings for too long, then the engines would be run up every seven days. It was not possible to do this whilst moored so the aircraft would be taxied, a job done by qualified aircrew only. Aircraft left in the water for three weeks or longer would be in a bad shape apparently.

The MAP and Farnborough had their input and the specifications changed as time went on, even RAF instruments and compasses were to join the conversion. Seven marks of the Catalina were processed, I, Ib, II, IIa, III, IVa and IVb, and the first amphibian Mk III arrived in April 1942. This model was part of an urgent programme, but when it was realised that the undercarriage chassis took up so much space and weighed so much that the equipment to make the conversion could not be fitted no more amphibian Catalinas were purchased.

In June 1943 decisions about the Shetland flying boat had to be made by the Air Ministry. The earlier work done by Saro was valuable and the company wanted some priority for further work to be carried out, with a view to post-war passenger flights. The Air Ministry decided to compromise with Saro – continue with construction, but henceforth the Shetland would be a transport aircraft only. One prototype was completed in 1944; DX166 made its maiden flight in December but the Air Ministry cancelled an order for production aircraft in April 1945. Subsequently DX166 was destroyed by fire after an accident in January 1946. The other prototype, DX171, was not completed until 1947 and was registered as civil transport (G-AGVD). It was scrapped in 1951 without carrying a single passenger and so ends the sorry story of the demise of the flying boat era.

Although this was not an RAF station the occasional visit was made by Sunderlands, whether for technical reasons or as a diversionary landing in poor weather, and it was considered to be a fine place to enter and leave its waters. This was to be proved when a series of flight tests were made at Beaumaris, with various marks of Spitfire fitted with floats. The idea behind this was to see if aircraft could operate in terrain where airfields could not be built. One Mk V was converted

The slipway today at Beaumaris. The road that passes the house is the B5109. (I. Jones)

Sunderlands were to be seen occasionally at Beaumaris to enable radar and signals equipment to be fitted. (D.J. Smith)

Spitfire Mk V experiments took place here on many occasions, to see how well it behaved with floats fitted. This occurred with a Mk I also, all with an eye for the Far East fight to come.

in April 1940, another in September 1941, two more in August 1942 and in the spring of 1944 a Mk IX (MJ982) was converted to a floatplane with a view to its use in the Pacific war theatre. Jeffrey Quill flew it for the first time on 18 June 1944 from the slipway at Fryars, but soon after testing began, the idea was dropped and MJ982 was re-converted to a landplane.

Also in 1944 a floatplane version of the Auster Mk V received testing, but was transferred to Helensburgh to complete. Another Auster, many years later in 1949, was converted for use in the Falklands and Antarctic. After testing it was shipped over there but was scrapped in 1955 because of corrosion.

In December 1944, when the threat from the U-boats had lessened, the possibility of using the Catalinas in other roles was considered. Air–sea rescue was one of the obvious roles and the company was asked to prepare a RATOG design that would make the flying boat suitable for this. A trial design was constructed and installed in the MAP DTD aircraft JX215 at Beaumaris. Also, early in 1945, the firm was asked to convert a squadron of Mk IVs, enabling them to carry an airborne lifeboat. With the approaching end of the war these ideas were dropped after a few trials.

Conversions to floats were also extended to the Auster 5 at Beaumaris in 1944.

HMS *Dark Aggressor*, one of five fast patrol boats of the Dark class, was built at Beaumaris in 1953–55. The company designed and built many light warships here.

The VR2 hangar, with extensions, was at the western limits of the facility. Note the side annexe. (I. Jones)

The north side of a B1 hangar, nearest to the main entrance to the site. It has endured three major extensions to its flanks. (I. Jones)

The B1 is at the top and the T2 on the right. The gradient is more severe than it appears in this shot. (I. Jones)

The B2 hangar at the north-east corner of Saro, showing the ubiquitous annexe on the far side. All in all this is a very interesting place. (I. Jones)

By the war's end the company had converted 300 Catalinas and when the last of these was delivered in 1945 the site was closed down. Its aviation days were over.

Saunders-Roe had a wealth of experience in the manufacture of boats pre-war, add to this its experience in forming aluminium and Duralamin, and at Fryars the company produced a multitude of Army bridging pontoons, bridging ferries and floating servicing platforms for flying boat bases. They made powerful, high-speed torpedo boats and minesweepers for the Navy, ships lifeboats and air–sea rescue lifeboats designed to be carried under the bomb bay of Shackleton aircraft. They made assault craft for the Army, and bus bodies that were to be fitted to Leyland chassis for export to New Zealand and other countries, as well as for London Transport, and all these items were in aluminium.

As these contracts ran out and the various buildings were sold off, large redundancies occurred, leaving just one factory to produce an undisclosed Navy contract. A couple of the hangars now appear to be used for warehouses but that is all. The site is deserted and now up for sale.

The hangars and buildings can only to be named using one's own interpretation, as the development of the site was ad-lib and urgent. There is no site plan, as there would be if it were an RAF facility, and every original hangar has been extended or had annexes added, sometimes annexes were added to annexes. Brickwork has been added to support hangar end walls and some really weird extensions have been made to the side of the T2 hangar. This T2 is famous for its work on naval boats and it is from this hangar that the rail track runs from 40ft above sea level down into deep water for launching. Aside from the T2, there is a B1, a B2, a VR2, numerous workshops and office blocks. What a great job was done here in this rural, thinly populated place, with a few craftsmen, quickly trained men and women, backed by a design and research team and the bosses who knew the business, all situated in a lovely scenic part of Wales.

This PRU photograph, dated 16 January 1946, shows the town of Denbigh, in particular the Howells School for Girls. The sports grounds belonging to this establishment were to become the cubstrip for the US Army unit based in the town. (Welsh Assembly Government Photographic Archive)

ELEVEN

The US Army Cubstrip at Denbigh

Grid reference SJ 059662

The British Army had used Denbigh as a base with hutted and tented accommodation to the north of the built-up area in the early years of the war, but the British made way for the Americans to use these camps in the lead-up to the invasion of France.

Among this influx of US Army units was the 733rd Field Artillery Regiment, which made camp on the southern side of the town, on the sports fields of the Howells School for Girls.

The regiment brought with them their establishment of Piper L4 Cubs as Air Observation Posts and it is believed that they used the flat area of grass adjoining the school fields as a landing ground. This area is now the Denbigh Town AFC's football pitch. The town absorbed the soldiers, doing their best to accommodate in every way and it is even claimed that artillery pieces were stored in the town's fire station.

The exact date of the Americans' arrival at Denbigh is not known, probably early in 1943, but we do know that they left in time to arrive on Omaha Beach on D-Day.

The Piper Cub L4 air observation post aircraft, as it was with the 733rd Field Artillery Battalion when based at Denbigh.

This is the plaque on the town hall wall, placed there to remind people of the occupation in 1943–44 of the 733rd FA Battalion. (D.J. Smith)

This composite photograph shows the sports field at Denbigh that was used as an airstrip by the US Army. (D.J. Smith)

This is the cubstrip at Denbigh today, looking east. (D.J. Smith)

Denbigh Fire Station today. This was used to store the artillery pieces of the 733rd FA Battalion. (D.J. Smith)

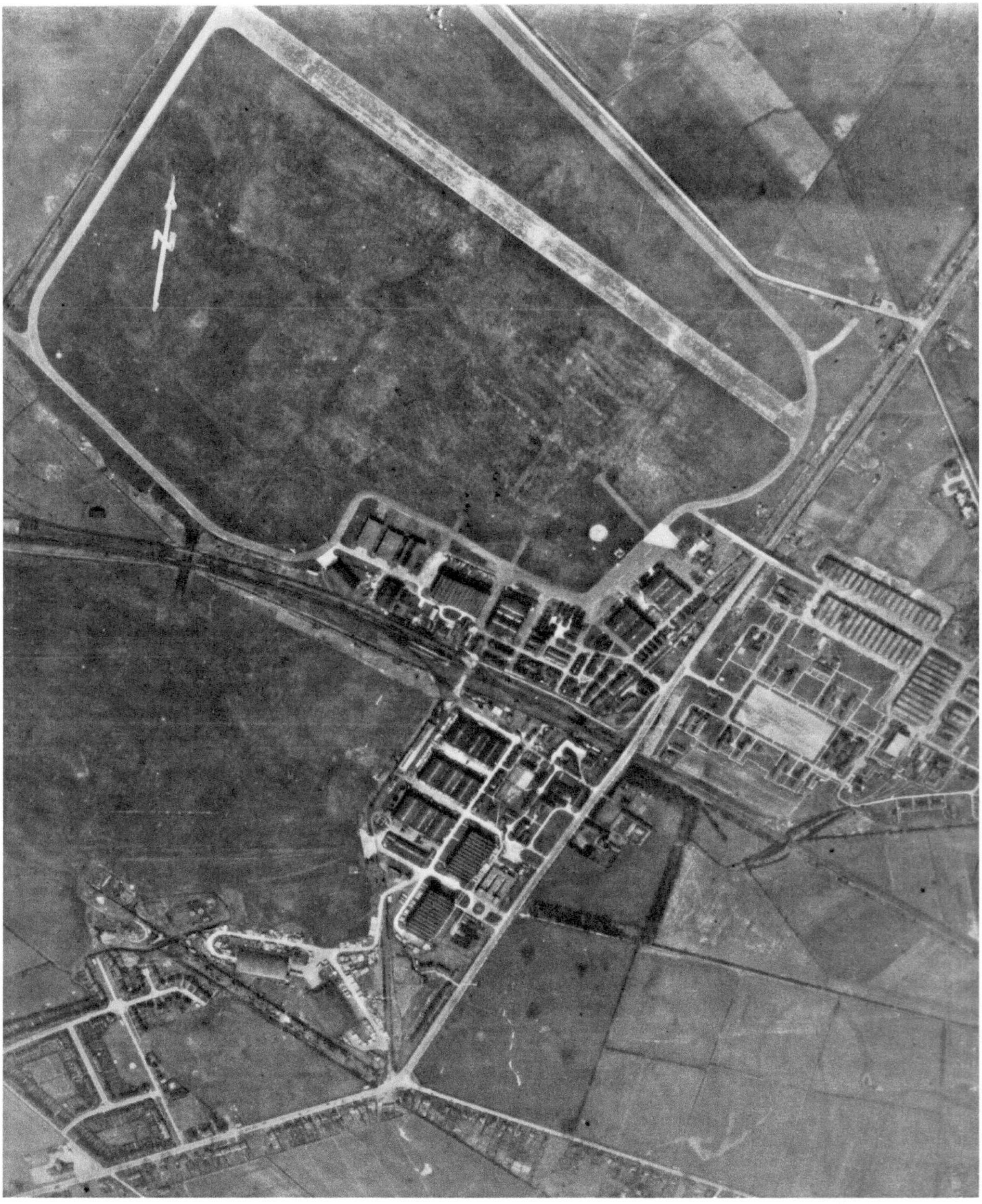

This overhead of 13 April 1947 shows the extent of the three parts of Sealand – East, West and South, with the sole hard runway on West Camp. (Welsh Assembly Government Photographic Archive)

TWELVE

RAF Sealand

Grid reference SJ 330700, 1 mile north-east of Queensferry, Flintshire, on the A550

The station of RAF Sealand was built on land reclaimed from the estuary of the river Dee, which had been canalised in the last century, saving the flood plains of the tidal area for land development. The original engineering work was carried out 200 years ago by Nathaniel Kindersley. The first road was built and bridges constructed to replace the ferries that were used at that time.

At the start of the First World War the area around Sealand was still a remote backwater with few buildings or people. Aviation was in its infancy and in 1916 a young engineer, Mr T. Murray Dutton, obtained an area of grassland on the edge of the river Dee and used it to start a flying school. He built a wooden aircraft shed, in which he decided to build his own aircraft to use in the school, mostly of the French Caudron warped wing type. The engines were second-hand and reconditioned on the site.

The Royal Flying Corps (RFC) was receiving its pilots from civilian schools at this time and after an inspection by officers of the RFC it was decided to take Mr Dutton's school on to contract. Shortly after the start of this contract it was decided to buy out Mr Dutton to make the aerodrome a full RFC station. The area of land that Mr Dutton's school had utilised was to be the area later known as South Camp, which represented only part of the total acreage required for this developing base. There was a lot of space available and two suitable stretches of land near Shotwick on each side of the railway line were chosen. The area north of the railway was to be named Shotwick, after the name of the nearest village, whilst the area to the south was to be named Queensferry. The latter was to be an Aircraft Acceptance Park, the former a training airfield.

Robert McAlpine moved in with his builders and accommodation for 839 servicemen was constructed, as well as six aeroplane sheds measuring 170ft x 100ft, built of brick walls and timber roofs. Offices on the airfield side were erected, along with a similar shed as an aeroplane repair shed. The landing area was roughly 1,000 yards x 900 yards.

Queensferry was built with three sets of hangars similar to those at Shotwick, but built in threes, measuring 300ft x 170ft. The two stations began with their own messes and accommodation sites, as it was expected that the Aircraft Acceptance Park would soon be receiving American aircraft from the docks for re-erection at Queensferry, but this never happened and no such deliveries were made before the First World War ended.

On 22 October 1917 No.37 Wing, Royal Flying Corps, had formed at Little Sutton, about 2 miles from Shotwick, and took control of Nos 90, 95 and 96 Squadrons, plus No.4 Training Depot Station (TDS). No.90 Squadron had formed at Shawbury on 8 October. No.95 formed at Tern Hill on the same day but moved to Shotwick on 30 October and was the first to arrive at the new airfield. No.96 Squadron moved in from South Carlton later that day and No.90 Squadron arrived in November.

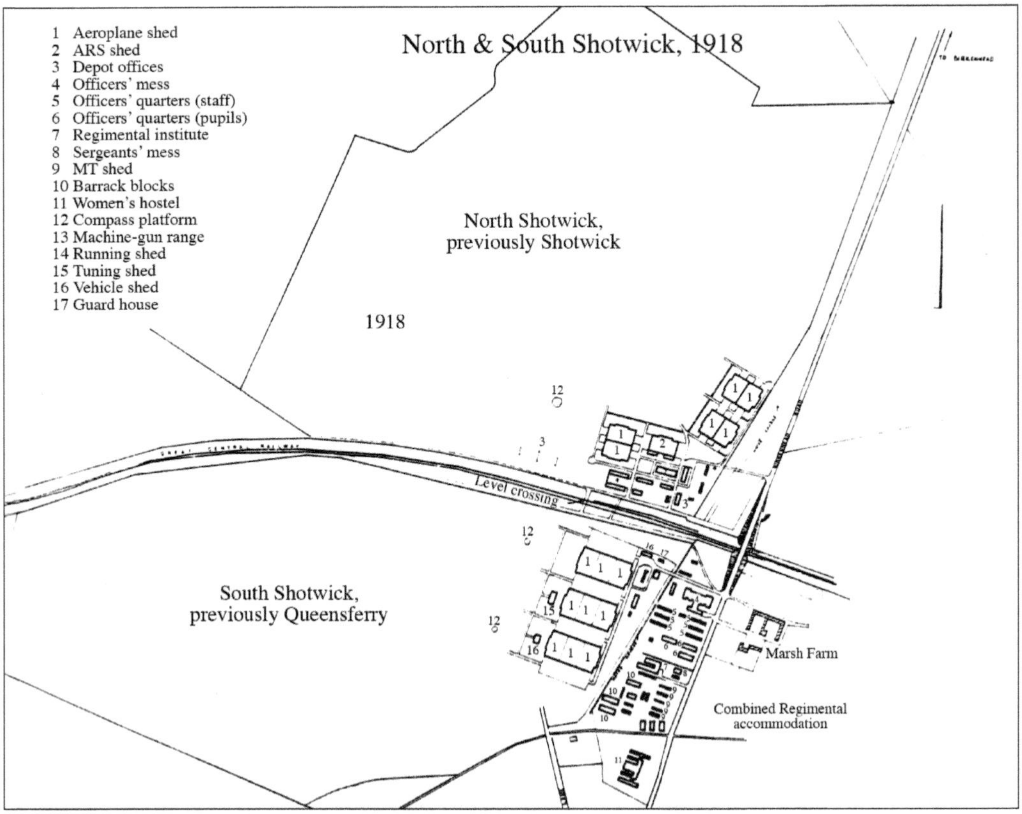

This map of 1918 shows the airfields of North and South Shotwick, as it was called then. There is no East Camp or hard runways yet.

This oblique shows West and South Camps in 1934. Little has changed since 1918.

The Sopwith Camel equipped No.67 Training Squadron at Sealand in 1918. (RAF Museum)

This Avro 504 was with No.5 FTS at RAF Sealand in 1920. (F. Jones)

The aircraft that could be seen at that time were Avro 504s, Dolphins and Sopwith Pups, which were always actively training when the weather permitted. Not only were the trainee pilots needed at the front but as soon as these three regular squadrons had been trained up they were to leave, making way for Shotwick to begin its long career as a major training station for the Royal Air Force.

The first of these training squadrons to form at Shotwick was No.67, which moved in from Shawbury on 1 April 1918 with Camels, Pups and 504s to take the novice pilots through basic training.

It was clear that Nos 95 and 96 Squadrons would not be ready for front-line service in time for active service because the war was ending, so they were removed for disbandment, making way for another training squadron. This was No.55 from Lilbourne, with Camels, SE 5s and Avro 504s, doing the same work as No.67 Squadron. These two units were to disband and be redesignated No.51 Training Depot Station, continuing the training of pilots.

At Queensferry very little was happening and nobody knew whether the RAF was to retain either of the stations in the peace time review, but squadrons were returning home to disband and many came to Shotwick.

No.4 Training School from Hooton Park arrived at Shotwick in 1919 as a separate unit and gradually took over the work of No.51 TDS, whose focus had changed from training to aircraft

Sopwith Snipe with No.5 FTS at Sealand in 1928.

This was the Avro 504N, the Lynx Avro, that was equipping No.5 FTS in 1928 at Sealand. (F. Jones)

A fresh batch of Avro Tutors that were to go on the strength of No.5 FTS, Sealand.

Armstrong Whitworth Siskin IIIs were to be seen at Sealand, along with the other trainer types, in February 1931. (F. Jones)

The Armstrong Whitworth Atlas trainer K2523 was introduced in 1934 and became 723M in October 1935 with No.5 FTS. (Phil Williams)

maintenance. The squadron used the hangars at South Shotwick, which was now the name of the site that had been Queensferry, for refits to aircraft and to store the huge numbers of redundant warplanes from the still-contracting Air Force. The TDS was so busy with this work that a Ferry Pool Flight was provided by No.3 Group to transfer the aircraft from Shotwick to other units in the UK. With this refitting work being done on the south airfield, South Shotwick had, at last, become a form of aircraft depot, even using Hooton Park briefly before its closure.

The reorganisation of the peacetime RAF continued and No.3 (Training) Group decided to open No.5 Flying Training School at Shotwick. In 1920 the new school opened with an establishment of Avro 504s. This new unit was to absorb No.4 Training School and No.51 TDS. The Commanding Officer of No.5 FTS was to be the CO of the two stations, but it was from the north airfield that No.5 FTS would fly.

The station was renamed Sealand in 1924 to bring an end to confusion with the RAF station at Scopwick in Lincolnshire. This new name, Sealand, was to unite the two sites of Shotwick and Queensferry as North Camp and South Camp of RAF Sealand.

Above: A Hawker Fury and a Tutor in one of the GS sheds at Sealand in 1937. (Derek Pratt)

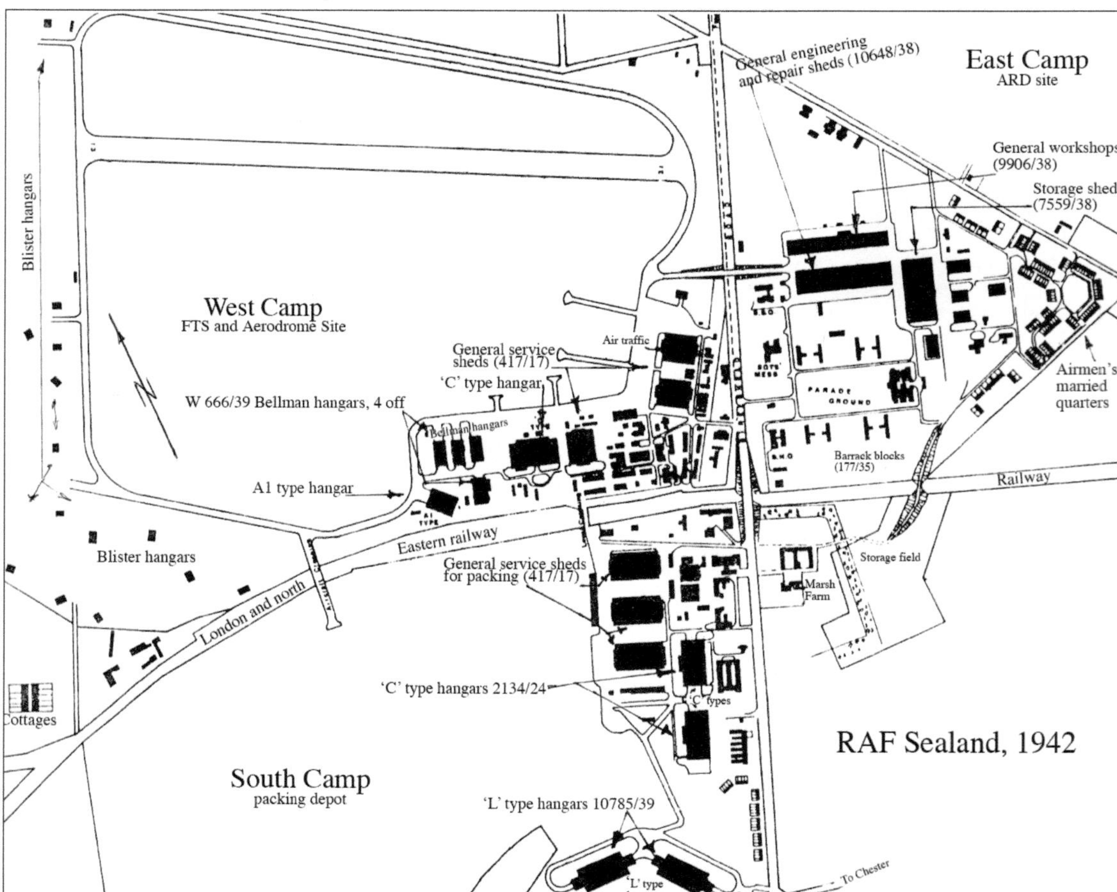

Right: This overhead shows some of the US Army's investment at Sealand, particularly the huge six-part stores and workshop building on South Camp (lower middle of picture). (Welsh Assembly Government Photographic Archive)

Opposite below: This map shows the station at Sealand at its highest development by the RAF in 1942. The Americans were to invest later in buildings, converting the airfield into a depot. (I. Jones)

An exciting statement of the base's retention was the arrival of the RAF Packing Depot from Ascot on 23 May 1929. This unit, based on South Camp, was to be responsible for the crating and dispatch of all the aircraft that had to be sent overseas.

The expansion at Sealand in 1927 provided better accommodation for its growing population. The establishment was to grow to eighty pupils and the old timber buildings gave way to permanent brick ones. The first of these was the officers' mess near the railway line on the south side. Next was the sergeants' mess, followed by five barrack blocks on the north side of the railway. The modern Hawker Tom Tit and Avro Tutor were the aircraft in use by No.5 FTS.

By January 1929 the Packing Depot had settled in and was busy receiving, dismantling, crating and dispatching aircraft around the world. The training of pilots continued, along with the regular accidents that occur with any flying training, but perhaps a little more here because of the problems of water on the landing ground. The water level was not far beneath the surface and with steady rain it became a worn, muddy morass, with nose in, tail up landings frequently occurring.

As the threat of another war appeared on the horizon, expansion caused the rate of pupil production to increase as there were to be 123 squadrons and a front line of 1,736 aircraft by 1939. To meet these requirements No.5 FTS, and others, had to change the system. Instead of 'initial' training at FTS the pupils would go to civilian schools for two months, then upon arrival at FTS they would receive intermediate and advanced training, which would include cross-country navigation flights and a period at armament practice camps. The old aircraft types that had been used for advanced training, such as the Siskin and Atlas, were to give way to the standardising on the Hart, Audax and Fury. The armament camps used were North Cotes Fitties and Sutton Bridge during 1936 and 1937, but when the newly opened Penrhos became available it was used almost all the time.

All the growth under the expansion scheme resulted in the formation of No.3 Aircraft Storage Unit at South Camp on 2 December 1935, holding aircraft before and after the packing unit had done its work, plus storage of other aircraft. By the time other storage units around the country were formed and given the title maintenance units (MUs) this one closed, absorbed into a larger packing depot and re-named No.36 MU.

On 24 November 1936 Colonel Charles Lindbergh was forced to land at Sealand after leaving Ireland on his round the world trip. He was forced down by fog which lasted for four days, during

A Miles Magister of No.5 FTS, Sealand, in 1939. (MAP)

which time he was the guest of the officers' mess and his aircraft was given a complete servicing by the mechanics of No.5 FTS. To commemorate the event and show his appreciation of the station's hospitality, he presented a silver rose bowl to the station, which still takes its proud place amongst the mess silver.

It was decided to introduce twin-engine training at the school in June 1938 and the first of the Oxfords arrived at No.5 FTS on 14 June 1938. Single-engine training was to be on the de Havilland Don, which was designed as a trainer, but proved unsatisfactory, so the schools continued with the Harts until the Harvard could be purchased and the Miles Master produced.

The packing unit was very busy as more squadrons were sent overseas. A serious thought occurred to the powers that be to move No.5 FTS to a new airfield and use Sealand as a vast maintenance unit. With the development of Maintenance Command there was a perceived need for aircraft repair depots and St Athan and Sealand would fit the bill, given their close position to high population and large ports for transportation. Plans were produced to develop the land to the east of the A550. It was requisitioned and what was to be named East Camp began its construction. The two stations, Sealand and St Athan, were planned as aircraft repair depots, with engine repair facilities and an aircraft storage unit in conjunction. The depots consisted of technical and domestic sites close together. The technical sites comprised the main engineering workshops, constructed on a steel frame with asbestos sheeting and measuring 620ft x 120ft; three 'C' type hangars, two of which were longer than normal at 400ft in length; a large steel and asbestos shed (store) 400ft x 200ft; an MT repair shed of 750ft x 20ft; timber, oil, paint and dope stores, station offices and an engine test house. Living accommodation was provided for an establishment of 1,350 personnel.

The station was taking the shape of the present-day station. South Camp was expanded to take two 'C' type hangars and other offices, and all the remaining First World War and later 1920s buildings were refurbished for their new role. Two more hangars ('L' types)were built on the southernmost side of South Camp as part of the aircraft storage unit and were absorbed into No.36 MU (the Packing Unit). The third 'C' type hangar was built adjacent to the railway on North Camp to supplement the three pairs of First World War general service hangars. These were later supplemented with no fewer than eighteen Blister hangars and an 'A1' type scattered around both airfields to provide additional cover for aircraft maintenance.

A recruit sub depot was formed on 1 June 1939 to fill a need until suitable places elsewhere were found, and it caused chaos. With no spare sleeping quarters and no clothing because No.30 MU was forming at the time, a tented camp was used until Padgate and Altrincham opened on 7 September 1939.

No.6 Salvage Unit formed at Sealand on 19 September in No.43 Group Maintenance Command. The unit collected its equipment from No.30 MU and the nucleus of the unit,

Miles Master Is and pupil pilots at No.5 FTS, Sealand in 1940. (Aldon Ferguson)

Station HQ offices at RAF Sealand. (Aldon Ferguson)

DH Tiger Moth of No.19 EFTS Sealand. This unit was here from 21 January 1941 until 31 December 1941. (F. Jones)

North American Harvards began to appear at Sealand in January 1941. They were the first to be shipped over in crates from the USA and were assembled here.

This is the slip-wing Hurricane. It was an attempt to gain more range for the aircraft, by using a biplane configuration until going into action, after aborting the top wing.

together with airmen and transport, plus stores, moved off to its base at Carluke, Lanarkshire, via No.14 MU at Carlisle, where it picked up more equipment. Its role, of course, was to visit crash sites, remove all salvageable material that could be used, then fill the hole created with debris and cover the site.

No.30 MU, which has already been mentioned, was another important addition to Sealand. It was formed here on 28 July 1939 as an aircraft repair depot, repairing engines, instruments, armament and wireless telegraphy. A planned aeroplane repair section (ARS) would only be formed after the rumoured departure of No.5 FTS became a fact. This new MU also had to wait for the buildings on the new East Camp to be completed. The three camps at Sealand were now to be known as West, East and South, not Shotwick or Queensferry.

On 1 June 1940 the first of the 'L'-type concrete hangars was taken over from the contractors and was immediately filled with three Lysanders, a Hind and two Battles, and on the same day No.36 MU was redesignated No.47 MU.

The Hurricanes of No.605 Squadron flew in to be packed and crated, to embark the squadron on HMS *Indomitable* for the voyage to Palembang in the Dutch East Indies.

No.5 FTS struggled in the winter of 1939–40. The wear and tear on this waterlogged grass airfield was holding up the production of much-needed pilots, so relief landing grounds were used at Speke, Hooton Park, Ringway and Borras (Wrexham). The construction of the airfield on a reclaimed estuary was called into question. It was all right for MUs but not for OTUs.

There were air raids of varying severity on Sealand. One, on 14 August 1940, did some damage, including a hit on the sergeants' mess which killed a wireless operator, six sergeants and six airmen from No.30 MU.

September 1940 saw the end of twin-engine training at Sealand, as the Miles Master was being delivered in numbers. These, with the occasional use of the Magister, were to serve No.5 FTS for a while. At this point there was an inflow of American pilots coming to aid Britain long before the US officially entered the war. Most of these men were at Sealand to do the initial course of training.

It was decided to move No.5 Flying Training School, to allow No.30 MU to gain space and expand its establishment, and it was to leave for Tern Hill by the end of 1940, twenty years after its formation at Sealand.

The work being done by No.30 MU was varied but vitally important and as soon as the FTS vacated the 'C' type and nearby Belfast hangars they were seized by No.30 MU. Bellman hangars were also promised. One of the projects undertaken by No.30 MU at this time was the fitting of A1 Radar to Beaufighters as soon as they were delivered by the manufacturers, as well as also converting seventy Havocs with Turbinlite equipment.

No.36 MU, the Packing Depot, was renumbered No.47 MU and was very busy. The pressure for more space was on it, too; for example, 700 aircraft were crated and dispatched overseas in February 1941 alone, including no less than 329 Fairey Battles to Canada.

The respite given to the airfield by the departure of No.5 FTS was short-lived, as No.19 Elementary Flying Training School (EFTS) arrived on 21 January 1941, bringing their Tiger Moths to this over-occupied station, dashing the hopes of the CO of No.30 MU that his 309 billeted-out airmen could adopt the barracks vacated by No.5 FTS. To make matters worse, the headquarters of No.81 Group formed here on 30 December 1940. The total strength at this time was 2,758, including 114 WAAF.

The first of the Harvards had been uncrated and erected by January 1941, marking the beginning of many thousands of these much-needed trainers. This month also saw the arrival of No.6 AACU (Anti-Aircraft Co-operation Unit) at Sealand with Lysanders, initially to serve the Rhyl and Chester areas in March 1941. By June this detachment had grown to one Lysander, one Dominie, two Dragons and two Leopard Moths, and their task was to exercise the searchlight and anti-aircraft guns of the north-east.

On 31 December No.19 EFTS was closed down. It had had a rough time on this busy station and the conditions, as with No.5 FTS before it, were not good for safety, resulting in frequent crashes. The school had also used Hooton Park, Little Sutton, Speke and Wrexham, and had trained ten courses, delivering 746 pilots.

Once again the breathing space was brief. No.24 EFTS moved in from Luton on 7 and 8 February 1942. This unit was to train Fleet Air Arm (FAA) pilots prior to joining the FAA's flying training schools and, ultimately, posting to the Fleet. This unit also flew the Tiger Moth.

The flying off and on of the heavier aircraft of Nos 30 MU and 6 AACU was tearing up the surface of West Camp and, at last, during the summer of 1942 the concrete single runway was laid, 3,960ft in length and 160ft in width. With a top surface of tar and sand it was to carry the numbers of 13/31 on its runway headings. The new runway was connected to the main aircraft servicing platforms and hangars by hard taxi-ways. No.30 MU was denied full use of the West Camp airfield by No.24 EFTS and they still needed more space.

In February 1942 eight officers and 134 airmen from the Polish Training Depot at Blackpool moved in to Sealand. They were the first of many who were to come here up until the end of the war. Initially they were shared out among all the sections of No.30 MU but by the end of 1942 they had been formed into a repair unit and were responsible for the repair of crashed aircraft, beginning with Mosquitoes.

In May there was a large increase in the numbers of WAAF, due to the completion of the Parachute Section, and on 29 June the first contingent of Polish WAAF arrived. The Engine Repair Squadron reached a record 105 Hercules engines in a month and the Lancaster Mk X programme was gaining momentum. Station strength reached a peak in November 1942, exceeding 4,000 for the first time, accommodated on the three main camps – East, West and South – with a WAAF hutted domestic site at Wepre Hall. There were 774 airwomen at No.30 MU alone and there were still 400 airmen billeted out!

An interesting sideshow performed by No.30 MU was the Slip Wing Hurricane. It needed a copy of the monoplane's wing fitted to the top of the Hurricane by struts that could be electrically separated in flight, reducing its biplane configuration to a monoplane one. A contractor, Mr F. Hills, was the inventor and trials were carried out at Sealand. The idea was to gain a longer range but this all fell through by 1943 as the jettisonable underwing tanks were developed.

No.30 MU's production continually broke its own record on consecutive months, yet it was still not up to full establishment. Besides airframes and engines, they dealt with clocks, watches, aircraft instruments, auto-controls, Sperry instruments, parachutes, harnesses, dinghies, tents, CO2 bottles, flotation bags, Link Trainers, generators, starters, magnetos, spark plugs, pumps, carburettors, WT receivers and transmitters, armatures, electrical instruments, engines, parts for engines, test rigs, balloon equipment and everything to do with motor transport.

No.24 EFTS continued its training both night and day. The school was split into four flights, A, B, C and D, with each course having two flights. Normally two would operate from Sealand and two from Little Sutton, all returning at night to Sealand. 24 August 1943 saw the first intake of twenty-four RAF cadets, who had been trained in America and were to be tested at Sealand for aptitude before leaving for operational training units. The school had fifty-one flying instructors and seven ground instructors by 1944.

From 28 April 1944 the school accepted navigators and bombadiers for flying training, but from now on the training of Royal Navy personnel ceased.

The school was instructed to move to Rochester, leaving Sealand on 11 March 1945, and they continued to operate from there until May 1946.

No.6 AACU's detachment expanded with more aircraft and pilots. Four Oxfords and two Hurricanes had joined them but had then disbanded on 22 November 1943 and amalgamated with elements of Nos 7 and 8 AACUs to form No.577 Squadron, with its HQ at Castle Bromwich. It left the same detachment at Sealand, which now held six Oxfords and three Hurricanes, plus one Stinson. They carried on their commitment to the needs of Army co-operation in the area and, additionally, in January 1944 this Sealand detachment of No.577 Squadron liaised with No.577's detachment at Bodorgan regarding the provision of targets over the Ty Croes range on the Anglesey coast, as Bodorgan could not meet its commitment. Most of the co-operation training remained the same – the Royal Artillery training regiment at Saighton Camp near Chester and other heavy AA units in North Wales as far south as Oswestry. A Master returned to the detachment in May and the detachment was now holding fourteen pilots, mostly Polish. Considerable co-operation was undertaken with the unit at Woodvale, No.285 Squadron, which had the same task but mainly operated to the north of the river Mersey. On 19 November the two squadrons swapped over – No.285 went to Sealand and No.577 to Woodvale.

At the beginning of 1945 No.30 MU was working on two main commitments in the Aeroplane Repair Squadron – breaking Lancaster Mk Xs down to produce and converting Spitfire Mk IX and XVIs to long-range standard. Other sections fully employed were: Armament Repair Section, Parachute Repair Section, Radio Repair and Manufacturing Section, Wireless Telegraphy Flight, Engine Accessory Shop, Woodworkers Shop, Motor Transport Shops, Engine Repair Section, Synthetic Trainer Repair Section, Radar Flight, Workshops Flight, General Engineering Squadron, Metal Workers Shop and No.5 Repairable Equipment Unit.

At the war's end the airfield was quiet and the run down began. By December the Lancaster programme was wound up, the last six Halifaxes had been converted and were awaiting delivery and all Mustangs were complete. The MUs still had work to do for units around the world, but similar MUs were closing down and No.47 MU was soon to be the only packing depot to survive the cuts that followed the ceasefire, just as it had been before the war.

No.30 MU was still busy but was frustrated by the demobilisation of many of the unit's craftsmen and specialists, and was finding it difficult to meet its many commitments as the main repairer of radar and radio equipment. It was then decided to make major changes to the product of No.30 MU. As the airfield and runway were small, and inadequate for the new and heavier aircraft of the jet age, it was to be motor transport and not aircraft that the MU repaired. The engine speciality was to change from aero to MT engines and in 1946 the unit was completely re-equipped to undertake the overhaul of heavy MT vehicles.

The Cold War brought the American Air Force back to the UK in 1948, and they were looking to stay. Their base at Burtonwood, near Sealand, had been a success in its wartime maintenance of UK-based flying units but, as big as this base was, they needed more. They wanted Sealand's accommodation blocks and the vast hangarage available. After many meetings it was decided that the units at Sealand could easily move on to allow for the massive occupation by the Americans. No.47 MU was now very small, with only 168 personnel still packing aircraft for overseas, so they were to move to nearby Hawarden to join No.48 MU which was already there. No.30 MU moved out to RAF Stoke Heath to amalgamate with No.24 MU, there doing similar work. They left on 15 March 1951, followed on the 31st by No.47 MU. The Americans had no use for the airfield at Sealand, as all their aircraft were to use Burtonwood, so the airfield at Sealand closed down on 6 March 1951, never to reopen.

The 30th Air Depot Wing USAF was to occupy Sealand. It was part of the 3rd Air Force and was allied to – but independent of – the 59th Air Depot Group at Burtonwood. The USAF used most of the buildings on all three sites at Sealand for the storage of goods for transportation throughout the UK, but before the end of the year this unit moved to Brize Norton, on 26 January 1952. This unit was replaced by the creation of the 7558th Air Depot Group (ADG). The mission of the new ADG was to operate the Sealand Air Depot, which would supply procurement, storage, distribution, evacuation, maintenance and disposal of USAF material, utilised by US Air Force and other US Government agencies of those areas designated to come under Sealand's responsibility. This was to be the whole of the UK, Norway and Denmark.

All this activity endured, until a decision was made in 1956 to vacate Sealand, along with many other US bases. The rundown commenced soon after and in January 1957 Colonel Wildrick, the CO, returned to the USA after three years at Sealand. The target date for the total withdrawal of US personnel was August 1957.

Coupled GS sheds on West Camp in 2008 (building No.48). (I. Jones)

This is a view from the north-west of General Service Shed (building No.111) showing the old Second World War anti-aircraft gun position on top of the brick-built hangar door abutment. These hangars are undergoing refurbishment. (I. Jones)

The typical barrack block at Sealand (building drawings 1424/26 and 1498/26). Thousands of lads passed through this type of accommodation during their training. (Aldon Ferguson)

On 18 July 7,000 tons of equipment started to arrive from all over the UK for disposal by local sale. The material was put on West Camp for storage and included tanks, howitzers, gun limbers, cranes, jeeps, tractors, charging sets and generating sets.

On 28 October 1958, after much debate, the RAF announced that Sealand would reopen in No.41 Group of Maintenance Command. In 1959 Sealand was to absorb MUs from Handforth, Stoke Heath and Hawarden, all of which were due to close. It was decided that West Camp was surplus to requirements and, therefore, reoccupation was to be at East Camp and South camp only, with the minimum use to be made of the old airfield site. No.30 MU reformed at Sealand on 1 February 1959, taking over all the work of No.24 MU at Stoke Heath. This unit had by now given up its airframe work and had become the major servicing and rectifying MU for airborne radar and radio equipment for the whole of the RAF at home and overseas.

During March a further move saw the closure of No.48 MU at Hawarden, which had merged with No.47 MU and was now responsible for the storage and packing of aircraft. This unit was to form the nucleus of No.2 MU, which re-formed at Sealand with the same tasks. Also in March No.3 (Electrical & Instrument) Squadron, formed within No.30 MU to carry out the tasks previously undertaken by No.290 MU from the now closed down Handforth.

By April the RAF had pretty much put all their eggs in one basket. The formation of No.30 MU was complete, as was the demise of Nos 24, 48 and 290 MUs. To complete the scene, a large detachment of No.61 MU moved into South Camp to join No.2 MU, the detachment coming from Handforth, and completed the make up of No.2 MU in the packing of aircraft and ancillary equipment for storage and transportation all over the world. The station strength on 30 June 1959 was twenty-four officers, fifty-six senior NCOs, 324 airmen, fifteen civilians of officer status and 493 civilians, totalling 912.

A further unit arrived that June when Maintenance Command Civilian School of Technical Training moved in from Aston Down. The transfer of equipment began on 25 June and was completed on 2 September. On 20 July The Air Ministry Works Directorate, based at Aston Down, also moved in to Sealand's new accommodation blocks, seeing the closure of yet another station at Aston Down.

The station grew with its technical skills and many changes were made over the following years. West Camp was sold off to a business man, Mr F.R. Ellis & Co., who developed the airfield site as an industrial estate in the late 1960s.

No.2 MU closed down on 30 June 1969; in spite of this the station strength continued to rise, reaching 2,072 by March 1971.

In July 1969 the accommodation on South Camp that had been occupied by No.2 MU was taken over by No.7 MU and became known as No.9 Sub. Site of No.7 MU (Quedgley), storing many items including crated aircraft parts and components, tents, Bessonau hangars and so on. This unit became No.6 Sub. Site of No.16 MU in December 1978 at Stafford.

Nowadays, No.631 Volunteer Gliding School flies from Sealand every weekend, subject to the weather, since the move from Hawarden in May 1963. Light aircraft and helicopters use the South Camp along with the gliders. The Headquarters of the North and West Region of the Air Training Corps and the headquarters of the Cheshire Wing are on the South Camp, together with the Technical Training School. This station has remained busy since its opening in 1917, and will probably be the last to close, as East Camp is still as important an RAF station as ever, under the guidance of the Defence Aviation Repair Agency.

Sites and Buildings

West Camp had six general service sheds in three groups of two, with an area of 170ft x 100ft. One aircraft repair shed ('C'-type hangar), four Bellman hangars, one A1 hangar, five over Blister, one double standard Blister and ten standard Blister hangars. A control tower was planned but apparently never built (drawing No.518/40).

There was a single runway of concrete, surfaced with tar and sand: north-west–south-east, 13/31, 3,960ft long x 160ft wide. In the early days of grass runways and flying from north and south airfields, the runways were: North site, QDM 310, 1,270 yards x 50 yards; north-east–south-west, 1,300 yards; east–west, 1,000 yards; South site, north–south, 1,330 yards; and east–west, 1,030 yards.

South Camp's buildings were as follows: nine GS sheds in three groups of three, two 'C' type hangars and two 'L' type storage hangars. The USAF, during its occupation, erected many fine buildings on this site and refurbished others of First World War-vintage, but the biggest and most impressive was building No.280, placed on the airfield of South Camp. It comprised six sheds, each larger than a 'C'-type hangar, of average bays, built as one building and complete with central heating. It was used as a vast store.

East Camp, the site of No.30 MU, had no hangars but many large workshops and stores, plus the usual RAF messes, institute, and barrack blocks.

This PRU overhead shows the station of RAF Hawarden on 17 January 1947. There are still many aircraft in the spaces between the runways, all awaiting the scrap man. The fields used as dispersals to the north-west are now empty, where once they were crammed. (Welsh Assembly Government Photographic Unit)

THIRTEEN
RAF Hawarden

Grid reference JS 350650, 3 miles south-west of Chester in Flintshire

When the decision was made to build shadow aircraft factories in the mid-1930s, several factors were considered as to where they should be built. Were there enough workers in the area? Was it handy to transport links? Most importantly, could you have an airfield on the site? Hawarden met these criteria. There was a large area of unobstructed land that had been reclaimed from the Dee estuary, known as Saltney Marsh, and the Government decided that this would be the ideal place to build a very large aircraft assembly factory for the production of Vickers Wellington bombers. Funded by the state, the project would also provide an airfield to test and deliver the Wellingtons, a maintenance unit (MU) and aircraft storage unit (ASU), an operational training unit (OTU) and along the way many other smaller units were to be absorbed.

Construction of the factory at the south-east part of the land began in 1937 by the contractor Sir William Arrol & Sons Ltd. It was to be a vast assembly hall where parts made at other factories would be brought here for final assembly. The plant took the name Broughton from the nearby town. The plant was managed by the agents of Messrs Vickers-Armstrong. The MU and OTU were built in the fields to the west. A part of this airfield site had been used by the aircraft of No.5 SFTS from RAF Sealand, which was only a few miles north of here. It was in use as a relief landing ground (RLG) under the name of Bowers, though was sometimes known as Bretton RLG.

The MU had the task of receiving the Wellingtons from across the airfield and fitted them up for service, up to RAF specification. The unit was formed on 6 March 1940, under the title No.48 MU. For this and other sites the building contractor was B. Sunley & Co. Ltd. An aircraft storage capacity of six 'L'-type aircraft sheds was constructed and as soon as they were ready the MU was swamped by their first quota of aircraft for storage. There were nine Lysanders, fourteen Wellingtons, one Magister, one Miles Falcon, seven Herefords, two Bothas and one Henley. The first of many Hurricanes arrived and twelve Lysanders were to be prepared for shipment to the Near East. These were fitted with long-range tanks and tropical equipment.

The first contract for Wellingtons to be placed here was for 750 in May 1939. This comprised three Mk Is, seventeen Mk Ias and, with further orders, 1,583 Mk Ics. The first Wellington to be produced at Broughton made its initial flight on 2 August 1939. It had been assembled in a Bellman hangar as the main factory was still incomplete. The OTU was swiftly organised – it was to be the only Spitfire OTU up until then – and the pilots were badly needed after the fall of France. The unit was formed here, designated No.7 OTU, and opened on 15 June 1940 with an establishment of twenty-five Spitfires (eight in reserve), fourteen Hurricanes (five in reserve) and thirteen Master Is (four in reserve).

In May 1940 'C' Section of No.3 Ferry Pilot Pool formed at Hawarden. It was the largest element of the Air Transport Auxiliary outside of its HQ at White Waltham. Its purpose here was to fly Wellingtons from Broughton's production to squadrons or MUs, fly aircraft to and from No.48 MU and the aircraft storage unit, and fly Blenheims off the production lines at Speke's shadow factory and Hampdens of English Electric at Preston. The unit had seventeen pilots and four Ansons.

OTU site No.4 (building No.225) PS tank, with pump house to its left (building No.233) and a pill box at front. These have all been demolished now. (D.J. Smith)

Hawarden's watch office/traffic control (drawing No.518/40 modified). (D.J. Smith)

The maiden flight of the first Hawarden-built Wellington L7770 on 2 August 1939. (Via D.J. Smith)

A Master Mk I and a Spitfire become involved at Hawarden's No.7 OTU. The Vickers factory is in the background. (Via D.J. Smith)

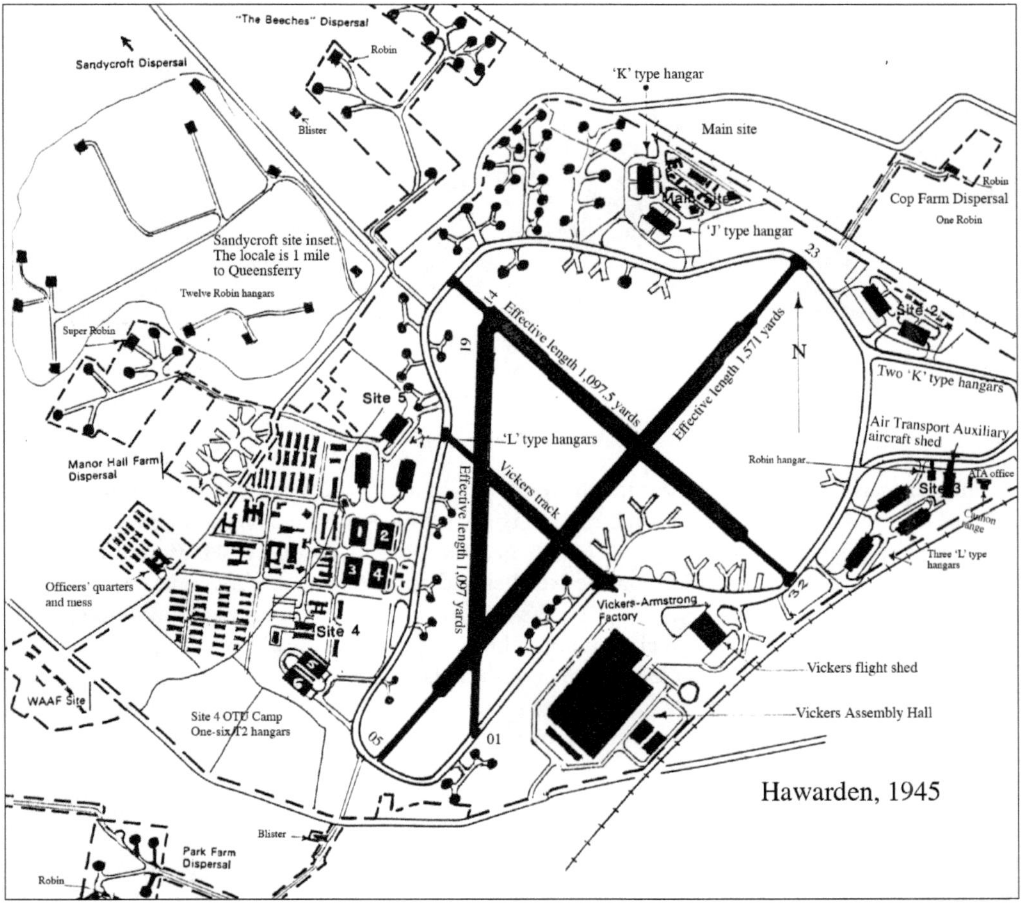

A map of Hawarden in 1945, at its final development by the RAF. (I. Jones)

The Vickers flight shed at Hawarden in 1944. (Via D.J. Smith)

A poor photograph but still precious, showing one of the last Mk I Spitfires to be built. It was on the strength of No.7 OTU, coded LV-N. This was used as a chaser in the tactical training mode. (Via J. Smith)

The previous system of training students from flying training and service flying training schools left the operational squadrons to teach the pilots the art of dog-fight survival at the war base of the unit. The operational squadrons no longer had the time to do this, so Hawarden had to do the job. This was always dependent on the weather and opportunity. Landings and take-offs were always competing with the airfield's use by the factory and the incomings and outgoings of the MU. The courses were of two weeks' duration, but if a pilot had his period of practice blocked by bad weather, he was likely to leave with just a few hours' combat training. The OTU was to administer the station with the MU as a lodger and at that time the station HQ was along with the OTU in a tented camp at No.5 site as there were still no buildings.

Hawarden was bombed by the Luftwaffe on 4 November 1940, destroying many aircraft. Sealand had been bombed earlier, on 14 August 1940. This caused concern, as Fighter Command's resources were stretched and were unable to put a unit in place to combat these attacks at that time. An unofficial battle flight was established at Hawarden to attempt to meet the bombers in any future raids on the north-west of England. Three Spitfires of the OTU were to be kept armed and fuelled, to be available to any of the instructors. Warning of the approach of the enemy was to come from the Observer Corps Headquarters in Manchester. This arrangement worked well enough to bring down an HE111 on Saltney, a JU 88 at Merioneth and a Dornier 215 into the sea, all during that late summer of 1940 when Great Britain really struggled.

On the wing of this Spitfire is Sgt T. Stivar, with a friend from New Zealand. Both were pupils with No.57 OTU in the spring of 1942. Stivar went on to No.310 Squadron and later, from 1944 onwards, he flew with No.13 Squadron. (Via D.J. Smith)

Above left: Cpl Jim Earnshaw, in the cockpit, and Cpl Hall, with Spitfire 1A-AR213, of No.57 OTU at Hawarden in the winter 1941–42. The red and white horizontal stripes on the cowling denoted this aircraft as an enemy 'jumper' aircraft. There were four codes allocated to No.57 OTU: LV, XO, JZ and PW(Jim Earnshaw, via D.J.Smith)

Above right: No.57 OTU instructor, P/O O. Hruby, with Spitfire I, X4926, LV-A, at Hawarden in 1942. (Via D.J. Smith)

Another 'enemy' attacker, Spitfire 1A-AR213, at No.57 OTU during the winter of 1941–42. (Phil Davies)

Fitter (engine) Jerome Geeson restores Wellington's pair of Hercules engines at No.48 MU, Hawarden, in 1945. (Via D.J. Smith)

In July 1940 No.7 OTU dispensed with its Hurricanes and by August the establishment was fifty-eight Spitfire Mks I and II, seventeen Masters and six Battles for target towing.

A 'Q' site at Borras was obtained, complete with a flarepath with lights on top of poles. This was, of course, before Borras became RAF Wrexham. There was also a decoy at Buckley, simulating a factory. Made of canvas and scaffolding, it looked effective from the air in daytime.

A crisis developed at the airfield during the winter of 1940–41. After enduring a famously wet winter, the place was flooded. At this time it was a grass airfield, though with all the movements it had become a morass. All the units and factory were behind in their work and people were decrying the fact that a marsh reclaimed from an estuary was not the place to build an airfield. Urgent deliveries of Wellingtons were delayed, the MU could not receive or dispatch, but worst of all the trainee fighter pilots could not get their hours in before going to combat squadrons.

The all-weather track that ran from the factory to the MU was used by some as a runway, although it was only 700 yards in length, and moves were made to extend this by a few yards and make it a concrete lifeline.

No.7 OTU was redesignated No.57 OTU on 28 December 1940 and became an all-Spitfire unit, equipped with Spitfires Ia, IIa, Vb and Vc, Fairy Battles, Masters Defiants, Martinets, Dominies and Proctors. Four codes were to be seen on the aircraft of this large unit: JZ, LV, PW and XO.

Mk 19 Avro Anson of No.173 Squadron. This unit was previously designated No.4 Home Feny Unit at Hawarden, from 1 February 1953 to disbandment in September 1957.

Mustang of No.41 OTU, Hawarden. (MAP)

Due to the poor conditions at Hawarden, all of the unit's flying took place at other stations nearby, such as Speke, Cranage and Borras, which was soon to become Wrexham. Two flights of Spitfires were detached to Sealand and one flight to Cranage in an attempt to get some training in.

Satellite landing grounds at Bodorgan and Tatton Park were acquired to store the spill over from the ASU and large tracts of land were taken over to the north, west and south of the MU, eventually developing six large dispersal sites equipped with roads, hardstands and Blister or Robin hangars.

When the airfield dried out in April 1941, the contractor – Gerrards of Manchester – moved in to lay the concrete runways that were so badly needed. One was of 1,600 yards and two of 1,100 yards, with taxi-ways and hardstands.

The 'B' Flight of No.4 Ferry Pilot Pool (FPP) RAF was also here from 11 October 1940, until being renumbered No.9 FPP and eventually disbanding on 1 April 1941. Taking its place came the formation of No.3 Delivery Flight, which was responsible for the delivery of aircraft from MUs to units within Nos 9 and 12 Groups. It was attached to the station of Hawarden, with six pilots and one Dominie. It stayed until January 1942 and then moved to High Ercall.

At the end of No.57 OTU's first year of operation it had suffered a total of sixty-four aircraft losses from accidents, such as taxiing into one another, landing on each other, hitting trees on approach and, most of all, mechanical failure. No.7 OTU lost five Spitfires in one afternoon on 7 December 1941.

Due to the needs of the Americans for bases and depots, all the personnel of No.37 MU at Burtonwood were to be taken on to the strength of No.48 MU at Hawarden, who gained not only a crowd of extra workers but also the No.37 MU's satellite landing ground at Knowsley Park.

The operational training unit at Hawarden had achieved the all-time record for the highest number of hours of flying in the summer of 1942, logging 5,282 during June. The station's aircraft movements in eleven hours on 6 September was 1,002 landings and take-offs. It was very busy, not only with the OTU but with Vickers, and the comings and goings from the MU and ASU caused friction; accidents were happening on the runways on a regular basis. The station had far too much traffic. The answer to the problem was soon to arrive in the form of the new airfield at Eshott, Northumberland, where No.57 OTU was posted off to in November 1942.

The MU at this time was still working on Wellingtons and Hurricanes, but the Halifax was being introduced. On 1 November 1942 Hawarden was transferred from No.70 Group to No.9 Group Fighter Command, and although the departure of 57 OTU had given the station a breather, it was not long before they would be back to crowding on the runways. No.41 OTU was due to arrive on 14 November 1942, bringing their Mustang Is. They were to teach Army co-operation but, after the disbandment of Army Co-operation Command, the teaching was to be the skills of fighter reconnaisance.

By now the satellite at Poulton had been constructed by Messrs Wimpy & Co. The fine new runways of what was now a relief landing ground for No.41 OTU were soon full of Harvard's Masters and Mustangs. The OTU, on the disbandment of Army Co-operation Command, had amassed a lot of pilots with little or no experience on fast single-seat fighter planes. These were men who had served in Lysanders and Blenheims and the transition was to be done at No.41 OTU. A small holding unit was formed within the OTU, at Poulton this time, named No.3 Tactical Exercise Unit to keep newly qualified, spare pilots in practice on Mustangs and Hurricanes, and later the Typhoon, while awaiting operational postings. This unit was disbanded during March 1944, but immediately reformed at Annan on 21 March 1944, with three squadrons of Typhoons. Two Seafire squadrons also arrived for tactical training, especially low-level reconnaissance. They were Nos 885 and 808 Squadrons, Fleet Air Arm, staying during September and October 1944.

More visitors arrived on 9 June 1944 when thirteen Halifaxes from Nos 420 and 425 Squadrons made landings at Hawarden, because all other landing grounds were fogged in, on returning from bombing French marshalling yards.

The winter of 1944–45 was severe, particularly January 1945 – the Battle of the Bulge period. Due to ice and snow it was impossible to fly aircraft in or out of the airfield, and by the end of the month there were 667 aircraft held here and at Hooton Park. The United States Army had more than 1,000 men transported to Normandy through this airfield during March 1945. On 16 July 1944 1,044 troops in fifty-nine C-47s and on 7 August 420 troops in twenty-one C-47s were transported to France. On 12 March 1945 another unit was formed at Hawarden, No.58 OTU, after part of No.41 OTU was redesigned. The Fighter/Reconnaissance Wing at Hawarden was to go to Chilbolton and retain its number, No.41 OTU. Then the Day Fighter Wing, located at Poulton, was to move over to Hawarden and become No.58 OTU. The move to Chilbolton included twenty-three Hurricanes, eleven Spitfires, six Masters, five Martinets and one Proctor. The re-formed No.58 OTU at Hawarden had seventy-five Spitfires, eighteen Masters and six Martinets. This OTU was then split into three squadrons – Nos 1 and 2 at Hawarden and No.3 at Poulton – for advanced training. Each pupil would spend six weeks at Hawarden and then move to Poulton for the final two weeks of the course. They practised battle formation, formation landings and take-offs, each pilot carrying out two low-level, one-hour, navigation flights, one duel in a Master and one solo in a Spitfire. These units were to disband a few months later on 20 July 1945, after a total of six courses having passed through.

At this period many allied personnel were posted to Hawarden to establish an Aircrew Holding Unit, pending their demobilisation or redeployment, and they were billeted in the vacated accommodation. When the OTU closed down Hawarden's station HQ passed over to the MU in Maintenance Command, from No.12 Group Fighter Command, and the MU was very busy.

No.58 OTU was formed at Hawarden on 12 March 1945. They used the Miles Master trainer, among many other types, as the war was nearly over.

Former ASU and MU 'K'-type hangar at Hawarden (building No.2), now in use by Raytheon Systems Ltd. (I. Jones)

This semi-detached house on the airfield was the HQ and crew room of the Air Transport Auxiliary in the first year of the war. (I. Jones)

On site No.2, this former No.48 MU 'K'-type hangar now has Raytheon Systems using its eastern end and Raytheon Aircraft Services at the western end. (I. Jones)

'L'-type hangar No.1 on site No. 3 at Hawarden. This former ASU and No.48 MU hangar is now used for Hawker fuselage construction. (I. Jones)

'K'-type hangar on the main site (building No. 1). It is now the home of Hawarden Air Services. Again, this was used by No.48 MU during the war. (I. Jones)

A privately owned Jet Provost on the ramp at Hawarden Air Services' hangar No.1, on the main RAF site. (I. Jones)

New hangars, all in private company or individual hands, such as the supermarket Iceland and the Duke of Westminster. This was the site of the panhandle dispersals on the main site in the north of Hawarden. (I. Jones)

Where No.48 MU's aircraft storage unit was concerned, they had more aircraft than they knew what to do with. Like many other airfields in the UK at the end of the European war, retired warplanes were stored on fields but farmers and other owners of these sometime-productive fields wanted them back for agriculture. At the end of June 1945 1,091 aircraft were in stock, including the sub-site at Hooton. This became 1,127 in July, and peaked at 1, 177 in September.

When the fields were given back to the farms, the aircraft had to be brought into the boundaries of the airfield proper, and the place was clogged with them awaiting scrapping. Some aircraft were to be refurbished and delivered for sale, mostly to countries in Europe to enable them to redevelop their own air forces. The MU still had a fair-sized commitment to supply new aircraft to the RAF.

With the voluntary Air Transport Auxiliary – in the form of No.3 Ferry Pilot Pool – vacating the station in November 1945 and civilian pilots returning to peacetime employment, it was replaced by No.4 Ferry Unit (RAF). This unit was formed at Hawarden on 30 November 1945, a large organisation equipped with Blenheims, Mosquitoes, Anson 19s, Beauforts, Halifaxes and Proctors. It was disbanded on 7 February 1952 to be redesignated No.4 Home Ferry Unit.

When RAF Atcham was closed in April 1946, the detachment of No.577 Squadron that had resided there was transferred to Hawarden. No.577 had formed at Castle Bromwich in December 1943 by merging elements of Nos 6, 7 and 8 Anti-Aircraft Co-operation Units, their anti-aircraft co-operation task continuing as before, with Western Command at Chester. Their equipment was mostly Oxfords, Hurricane IVs and Spitfire 16s for high-level radar flights for No.211 Anti-Aircraft Regiment. They also had Vengeance TT aircraft. This detachment disbanded at Hawarden on 15 June 1946.

Mosquito Till TT 299 outside the flight shed at Hawarden in July 1963. It is, of course, a de Havilland plant at Hawarden now.

De Havilland's factory was very busy during the 1950s and '60s. This view shows some of their products at that time: Hornets, Chipmunks, Venoms and Vampires.

For some time the Vickers plant had been running down its Wellington work, after producing 5,540 of them, and had been producing the Lancaster. They delivered 235 of these before production ceased in September 1945. The company was able to keep most of its workers busy by making aluminium prefabricated houses – at least for a while, as the plant was to be taken over by de Havilland to help the company produce aircraft to meet orders for the Doves, Herons, Vampires, Hornets and Mosquitoes that Hatfield, the parent factory could not cope with.

No.4 (Home) Ferry Unit disbanded at Hawarden on 1 February 1953 and was redesignated No.173 Squadron, with Ansons XII, XIX, Varsity T1s and Venom NF3s.

The MU continued scrapping aircraft, Halifaxes, Wellingtons and Ansons, until September 1957 when the No.173 Squadron was disbanded. The MU was then also closed down, on 2 September 1957, at Hawarden.

A further small MU, No.47, had arrived in 1951 from Sealand as an aircraft packing unit and RAF Hawarden closed on 31 March 1959. No.631 Gliding School operated from here until moving to Sealand in May 1963.

De Havilland inherited the station's buildings and many attempts were made to make Hawarden a civil airport. Cambrian Airways were involved in plans to run services such as a Liverpool to Cardiff Hawarden link. Starways and British Eagle added the stop here on their Liverpool to London flights in 1963, but when the company wanted the runway extended to suit their intended purchase of BAC 111s it met the financial stumbling block that was common to all the regional pre-war municipal airports.

Hawarden's control tower, much as it was when built, but with the glasshouse added and much more sophisticated gear inside. The left-hand side of the perimeter track at this point was the OTU site. (I. Jones)

On 6 December 1977 a new company, Air Wales, began a Cardiff to Hawarden service with a nine-seater Navajo Chieftain. This was the first schedule flight through Hawarden for a long time, but it was short-lived as they only achieved a break-even figure. During June 1979 Air Anglia bought out Air Wales while the company was still flying the Cardiff–Hawarden route. The service was suspended and evaluated, but it failed the test. The last flight was made from Hawarden to Cardiff on 25 June 1979.

De Havilland's factory continued its success at acquiring orders from overseas as well as home, making Venoms, Vampires, Chipmunks, Lincolns, Shackletons, Comets, Nimrods and DH 125s, later to be HS 125s, on the company's consolidation of the aircraft industry into Hawker-Siddeley. The factory is, of course, part of British Aerospace now and is a major player in the Airbus programme, building the wings for these aircraft. There is a local flying club and a steady movement in executive aircraft.

The Airfield and Dispersed Sites

The main site was at the northern corner with one 'K' and one 'J' type hangar, as well as the usual technical site facilities: main stores, main workshops, headquarters offices and so on.

Site No.2's storage comprised two type 'K' hangars (drawing No.3084/39) and other buildings. Site No.3 comprised three 'L' type hangars, an aircraft shed for the ATA of an unspecified type and one Robin hangar. Nos 2 and 3 Sites were placed at the north-east corner of the airfield. Site No.4, the OTU site, was situated at the south-west of the station and carried six T2-type hangars and a substantial number of other buildings. Site No.5, the MU, had three 'L'-type hangars and many other buildings, all to the west of the airfield.

Park Farm dispersed site had one Blister and one Super Robin hangar. Manor Hall dispersed site had two Super Robin hangars. The Beeches had one Blister and one Super Robin. Sandycroft had twelve Super Robin hangars. Cop Farm had one Standard Robin hangar.

The control tower was built to drawing No.518/40 and has since had a visual control room added to the roof.

The runways were: north-east–south-west, 05/23, 1,600 yards; north–south, 01/19, 1,100 yards; north-west–south-east, 14/32, 1,100 yards. All were 50 yards wide and of concrete, wood chips and tar. The Vickers Track had an unofficial potential as a runway and had been used as such, but of only approximately 700 yards, if the wind was favourable! There were 100 hardstandings around the airfield's perimeter and approximately fifty more at the dispersed sites.

At the war's end the personnel figures were: 304 officers, 352 NCOs, 1,632 other ranks and 364 WAAF, totalling 2,652.

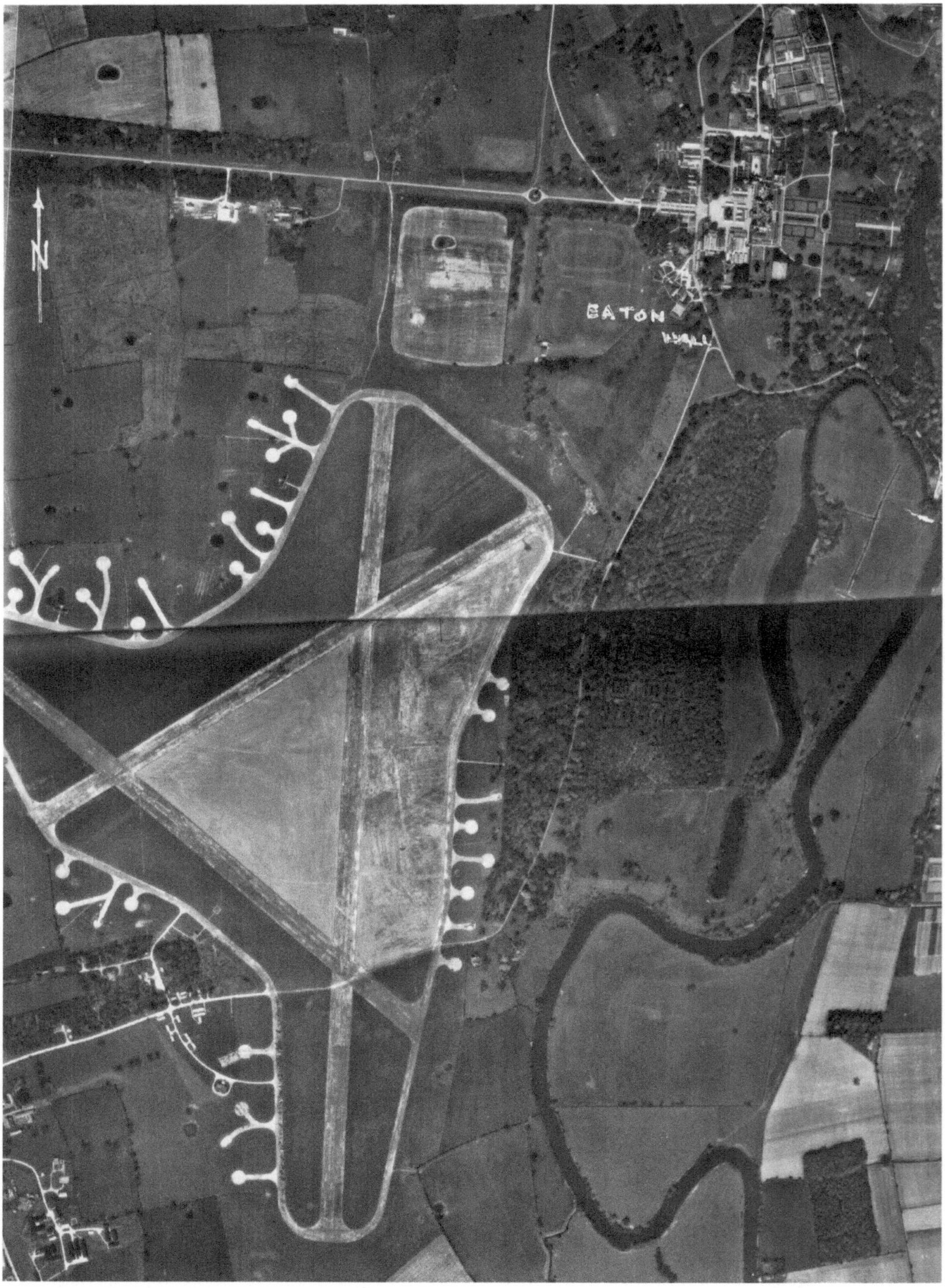

This composite overhead shows not only the airfield of RAF Poulton but also the old Eaton Hall, on 11 May 1951. The winding river that runs north to south in this picture is the Dee. (Welsh Assembly Government Photographic Archive)

FOURTEEN
RAF Poulton

Grid reference SJ 400600, 1 mile east of Pulford and 1 mile west of Aldford on the B5130 in Cheshire

This station was built late into the war on the land of the Duke of Westminster's Grosvenor Estate's Eaton Hall, by the builder George Wimpey & Co., as a satellite of RAF Hawarden, which lies some 4 or 5 miles north-west of the satellite. Hawarden had suffered from extreme congestion since it had been built and the opening of Poulton was sorely awaited. This occurred on 1 March 1943 in Flying Training Command, under No.12 Group Watnall.

The Hawarden-based No.41 Operational Training Unit (OTU) transferred its No.3 Squadron there almost immediately; more units would have gone with them but for the lack of accommodation, as the buildings were not all completed. Serious thought was given at this time to the use of RAF Sealand as a second satellite, but it was found unsuitable.

On 1 June 1943 Army Co-operation Command was disbanded and all of its units were transferred to Fighter Command. No.41 OTU was to adopt some of these ex-Army co-operation units and their training syllabus was to stay as it was, including ground support techniques and photographic reconnaissance. Part of the instruction was on Salisbury Plain at Larkhill, in order to give instruction with guns, artillery reconnaissance and observation of ground and troop movements. The base, when on these detachments, was at Andover.

No.41 OTU was equipped with Mustang Is, Hurricanes I and II and Harvards. It was swollen with the many aircrew left unseated by Army Co-operation's disbandment and was sent to Poulton to be absorbed into the OTU's system. Many of the airmen had flown only the slower aircraft such as the Lysander or Blenheim and there were quite a few accidents at Poulton, on the ground as well as in the air. On 14 August 1943 P/O Varley was involved in a prang here. Landing heavily in Mustang AG385, he bounced and attempted to go round again, then stalled with full power and flaps down. The port wing dropped and dug into the ground, causing the Mustang to cartwheel twice and come to rest in three separate pieces, the cockpit and fuselage up on the side of a Blister hangar, and the engine 30 yards away. The pilot suffered only slight abrasions.

On 21 June, Whitley EB402 of No.81 OTU, based at Tilstock, suffered an engine fire. Three of the crew bailed out; two landed without injury but the third fell in the river Dee nearby and drowned before anyone could reach him. The pilot and the other two crew members stayed in the aircraft and made an emergency landing; the undercarriage was retracted to avoid overrunning the runway. After the crew had got clear the Whitley caught fire and was completely destroyed.

On 1 November 1943, No.1515 Beam Approach Training (BAT) Flight arrived from Swanton Morley as a lodger unit, with a cool welcome awaiting them at Poulton where accommodation was already being fought over. They were upset to find that, with eight Oxfords on charge, there was no hangar, office or workshops for them.

It was a slow process for No.1515 BAT Flight. Even when the installation of the Standard Beam Approach equipment had to go in, there were delays. A Bessonnau hangar was found and

Hawker Hurricane of No.41 Operational Training Unit (OTU), which was later to become No.3 Tactical Exercise Unit (TEU) at Poulton. (F. Jones)

North American Harvards were used as trainers in great numbers at Hawarden and Poulton with No.41 OTU and later No.3 TEU.

No.1515 Beam Approach Training Flight arrived at Poulton on 1 November 1943 and left on 30 January 1945. The flight was never really welcome at this station because of overcrowding. They did most of their work at night, when nobody else needed the runways, using the Airspeed Oxford.

erected for them, and by the time their spell at Poulton was over on 30 January 1945, when they left for Peplow, they had trained many hundreds of pilots on short detachments from the parent unit – No.21 (P) AFU at Wheaton Aston. The BAT Flight was not a nuisance in their time here, as most of its flying was done in bad weather when the Mustangs were grounded.

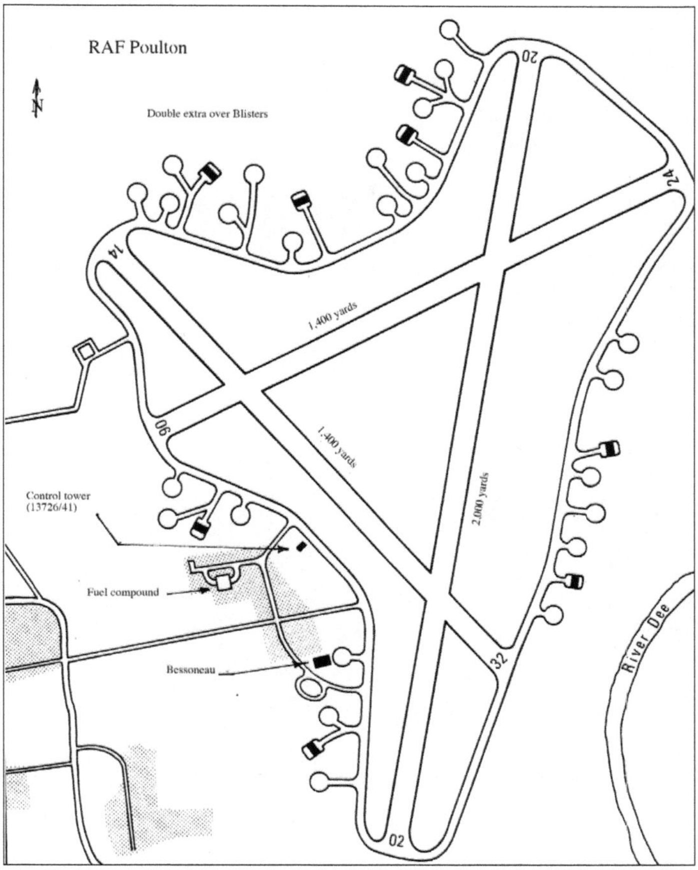

Map of Poulton illustrating the lack of buildings and hangars so typical of late-war satellite stations. (I. Jones)

At the end of November 1943, No.41 OTU underwent a thorough reorganisation due to an unexpected surplus of fighter-reconnaissance pilots. Each course output of pilots, starting with No.83 course, was not to be sent to operational squadrons but to stay at Hawarden and form a holding unit within No.41 OTU, to be known as No.3 Tactical Exercise Unit (TEU). The training syllabus of this small new unit was to be divided up between advanced gunnery, navigation and photography, and Army support.

No.83 course of No.41 OTU passed out on 30 November 1943, and on this day the TEU was formed at Poulton. The next three courses were to join the TEU as soon as they had passed through the OTU but then intakes to the No.41 OTU would be suspended until further notice.

The TEU was to comprise three separate squadrons by the end of December – Mustangs, Hurricanes and Harvards. This was to be somewhat rearranged when No.9 Group informed No.41 OTU that they must release thirty-four Mustangs in the near future without any guarantee that they would be replaced very quickly. The Hawarden/Poulton-based Mustangs were the Mk I, Allison-engined version. The Merlin-engine pilots were trained at Rednal by No.61 OTU in the skills of air fighting alone.

The Mustangs to be withdrawn from the TEU were to be selected from those with the least flying hours flown and were to be serviced and prepared for operations. Any that were not fit had to be made so before the end of December. None of the selected were to be flown but to be stored at Hawarden's hangars until collected in due course, as reserves for the coming invasion.

Further changes took place on 23 January 1944, at the OTU. In order to relieve congestion at Hawarden, all the Harvards, ex-'A' Flight, were to move to Poulton to join those already there, in order to make use of the accommodation that was coming on stream at the satellite and to standardise the maintenance of the Harvard at that base. All the Hurricanes of 'A' and 'B' Flights

Above and top: These two photographs of the Mustang crash were taken for the official investigation into its cause. (Via the Eaton Hall Archive)

Wing Commander David Annand, CO of No.41 OTU, seated in his Mustang at Poulton. (Via D.J. Smith)

were to be distributed between 'C', 'D' and 'E' Flights at Poulton. Four Mustangs of 'E' Flight at Poulton were to move to 'A' Flight at Hawarden.

At the beginning of March 1944 the aircraft on strength consisted of forty-four Mustangs, thirty-five Hurricane Is, four Hurricane Ias, two Hurricane Ics, one Hurricane IV, fourteen Harvards and ten Martinets. On 6 March No.86 course returned from leave and were all transferred to Poulton, to be formed into Nos 1 and 2 Squadrons within the Tactical Exercise Unit. Two weeks later, on 21 March, No.41 OTU resumed its intake of pupils. The training plan was for courses of twenty-two pilots to commence at three-week intervals. The duration of each course was nine weeks, plus a week's detachment to Andover for training in artillery reconnaissance with Hurricanes.

Aircraft strength was now inadequate to cater for the expanding TEU, so the commanding officer of No.41 OTU, Group Captain Alloway, complained to Group that it was impracticable to expect his unit – No.41 (OTU) – to be run in combination with the TEU at the same aerodrome, as there was neither accommodation nor the instructors available. The situation was aggravated further by the arrival at Poulton of the Advanced Flying Unit from Spitalgate. This was a detachment of No.12 (P) AFU and its base was at Grantham (Spitalgate). This detachment

Mustang III, FR 408, under repair while with the tactical exercise unit at Poulton in 1944.

Blenheim Mk V, DJ702, of No.12 (P)AFU at Poulton. (Aviation Bookshop)

Another Blenheim V of No.12 (P)AFU. This aircraft served with the Woodvale detachment at Poulton in February–March 1944. (D.J. Smith)

Poulton watch office/control tower (drawing No.13726/41). Systematic demolition twelve years ago has seen the last of this building, along with all the others. Luckily the late Barry Abraham took this picture before it came down.

had been operating from Woodvale and was tasked with advanced training on Blenheim Mks IV and V, to bring pilots up to the entry to Beaufighter operational training units.

Grantham's grass field was badly cut up from heavy usage and relief landing grounds were used, such as Woodvale, while steel mesh runways and coir matting were laid. The Blenheims of the detachment had only been at Woodvale for a month when a requirement arose for the airfield to house a fighter wing for a short period, forcing the move to Poulton.

All the accommodation at Poulton was spoken for, but two Nissen huts and a Blister hangar were grudgingly allocated. Two more Blisters were provided later. The Blenheims were only to be here for six weeks as it happened, and went back to Woodvale on 21 March 1944.

No.41 OTU and the expanding TEU continued their training. In the cold weather of January 1945 flying virtually stopped at the two airfields. The snow and ice played havoc with training and No.41 OTU's first accident of 1945 involved Hurricane LF 372, abandoned by his pilot near Whitchurch on 21 January. Spitfires were joining the OTU now and AD557 and BL 367 both made wheels-up landings, owing to technical difficulties. No.3 TEU began receiving the Hawker Typhoon Ib in numbers, but the protests of the CO that two training units on the base was not controllable bore fruit. It was decided to post off the surplus pilots to holding units under the 2nd Tactical Air Force, which became known as Group Support Units. This was done and No.3 TEU closed on 28 March 1944, although the number had been taken up again when No.4 TEU at Annan was redesignated No.3 TEU on 24 March 1944. At the time of closing down the TEU had changed its establishment, with its accent on Mustangs, to a heavy preponderance of Typhoon Ib. The (No.1) Squadron had a squadron of Typhoon Ibs, coded UW, and also a squadron including Masters, Mustang IIIs and Martinets, coded ZX. The (No.2) Squadron contained a squadron of Typhoon Ibs, coded EH, and a squadron of Hurricane IIs and IVs, Masters, Martinets, Mustang IIIs, Oxfords and Dragon Rapides, coded PA.

The final reorganisation took place in March 1945, when No.41 OTU was posted off to Chilbolton in Hampshire, retaining its unit number. They took with them twenty-three Hurricanes, eleven Spitfires, six Masters, five Martinets and one Proctor. The day-fighter wing that had built up at Poulton was to be redesignated No.58 OTU on 12 March 1945, from an element of No.41 OTU. The new OTU was divided into three squadrons, Nos 1 and 2 at the parent and No.3 at the satellite, for advanced training. Each pupil would spend six weeks at Hawarden and the final two weeks of the course at Poulton. During this final two weeks each pilot would drop a total of sixteen bombs on the Fenns Moss Range, near Whitchurch, and carry out air-to-ground firing at the Prestatyn Ranges on the North Wales coast. Emphasis was placed on observation and each flight of navigation training had to report on convoys, train movements and any other aircraft spotted in the air during his flight. The object, of course, was to make the trainee keep a good

No.3 TEU. No.2 Squadron took on a number of Hawker Typhoons before being absorbed into No.41 OTU.

No.58 OTU was formed from an element of No.41 OTU on 12 March 1945. This is a Spitfire of No.3 Squadron of the new unit.

lookout, as a reconnaissance pilot should. Sometimes an instructor would tail the pupil, in another aircraft a long way behind, to check his report on observations with that of the trainees.

The accident rate was very much lower than in the early years of training, but there were two serious crashes, both involving Poles. On 29 May Sgt Bauerek crashed in Spitfire BL 688 near Bury, probably due to getting lost in cloud and descending to find his position. The following day Sgt W. Polowancywk, in Spitfire NH 437, made a dummy attack on an RAF lorry on the A41 near Malpas, but misjudged and hit it. The Spitfire ploughed across a field and the pilot, although seriously injured, survived. Two airmen in the lorry were killed.

As the end of the war approached, the training of pilots was slowed down. Some of the OTUs were closed before the end of European hostilities. The turn of No.58 OTU came on 20 July 1945. With the disbandment of the OTU, the station HQs of Hawarden and Poulton were taken over by No.48 MU while things were sorted out as to the future of the two airfields. Poulton was not needed and reverted to a state of Care and Maintenance, followed by derequisition later.

The runways were used for a time post-war by the Duke of Westminster's private aircraft, but not any more as the duke uses Hawarden airfield now. A helicopter pad is adjacent to Eaton

A view from the northern end of the main runway, looking south. The north-west–south-east runway crosses this at the top of the rise, in the far distance. The south-west–north-east runway crosses the main runway at the foot of the rise. (I. Jones)

This is the southern end of the main runway at Poulton. It has a new farm built on it now. (I. Jones)

Belgrave Drive, one of the lovely entrances to the estate. The spire and hall are ahead, with the memorial between. The airfield is to the right of the picture, the most northern part some distance away. (I. Jones)

The same place as the over Blister hangar photographed in the Mustang incident. The line of young trees obscures the spire of Eaton Hall but it would be roughly where marked. The position of the workmen is also marked on the grass. (I. Jones)

The north-east end of the north-east–south-west runway is merely a track for estate vehicles now. The spire of Eaton Hall can be seen on the left. (I. Jones)

Hall and the old airfield. Eaton Hall had been commandeered during the war by the Army, then Dartmouth, the Royal Naval College, was bombed and the greater part of it was evacuated to Eaton Hall and its estate. The 1950s saw it occupied by the Army again, as an officer training school for men doing their National Service. When this property was derequisitioned, it was in such a state that it was demolished and a finer house built.

The airfield

The airfield had three runways of concrete and wood chips: north–south, 20/02, 2,000 yards x 50 yards; north-east–south-west, 24/06, 1,400 yards x 50 yards; north-west–south east, 14/32, 1,400 yards x 50 yards.

There was one Bessonau hangar on the south-west perimeter technical site and sixteen Extra Over Blister hangars (drawing No.6A/55/42), built in pairs on eight frying pan hardstandings around the perimeter. The control tower/watch office (drawing No.13726/41) was on the western side of the airfield, at the technical site. There were five other sites, No.2 Communal, No.3 Sleeping, No.4 Sleeping, No.5 WAAF and No.6 Sick Quarters. Personnel at the end of the Second World War totalled 1,077, comprising: seventy-nine (RAF) and two (WAAF) officers; 178 (RAF) and eight (WAAF) senior NCOs; 720 (RAF) and ninety (WAAF) other ranks.

Archaeological digs are taking place around the runways, indeed even under them. Apparently finds of Stone-, Bronze- and Roman-age periods have been found in this area, and a great interest in this private estate has been awakened.

RAF Wrexham is known to the locals as Borras Aerodrome. This RAF PRU photograph is dated 1947. (Welsh Assembly Government Photographic Archive)

FIFTEEN
RAF Wrexham

Grid reference SJ 365525, 3 miles north-east of Clwyd on the A534

Long before becoming an RAF station, the land around Wrexham had seen lots of flying. Land further away at the racecourse had been used extensively for flying before the First World War, when Gustav Hamel entertained the public in the displays of August 1912 and June 1913. Wrexham Borough Council were talking of making the racecourse the municipal airport, but it was the area 3 miles out of town to the north-east that was to develop, firstly as a civil flying field, then having casual visits by aircraft of RAF training schools, and lastly as a Royal Air Force Station in Fighter Command, then Flying Training Command.

Wrexham Rural District Council had permitted and encouraged flying at Borras Lodge because of its use by the aircraft of, firstly, No.4 Training Squadron, RFC, flying from its base at Shotwick from 1917 until the end of the First World War. Secondly, No.51 Training School flew from Hooton Park until 1920 for circuits and bumps, and to practise emergency landings.

These grass fields were used from just after the First World War for private flying and were collectively known as Borras Aerodrome, after the lodge of that name nearby. Lancashire Aero Club and Liverpool & District Aero Club used these fields on weekends.

In 1932 and again in 1933 Sir Alan Cobham brought his National Aviation Day Circus to Borras. The Charles Scott's British Hospital Air Pageant came in 1934 and the Jubilee Air Display in 1935. From 1936 it happened annually, until the Second World War, under the name of the British Empire Air Display.

All of this aerial activity had not been ignored by the Air Ministry, or those people responsible for finding suitable sites for military airfields in the face of the looming war, and notes were made. When the war started the aircraft of the training bases nearby, like Sealand and Hawarden, were using Borras but it was not yet an official Royal Air Force airfield.

The fields were on a high plateau and were well drained, unlike Sealand where most of the Hawker Harts and Audaxes came from. That station was suffering from the problems of putting an airfield on land reclaimed from the Dee estuary and then adding hundreds of aircraft movements each day, with a very wet winter. Worse was to come, as the building of RAF Hawarden was also on wet, reclaimed land from the Saltney Marsh and also had a training commitment.

By February 1940 No.5 Service Flying Training School (SFTS) was using Borras as a relief landing ground all the time, simply because it was dry, except for some puddles in one corner. Masters and Oxfords flew in, replacing the old biplane trainers and then minimal efforts by the RAF to prepare a decent landing ground were made. Hedges were removed, and three grass runways were rolled: west-north-west–east-south-east; south-west–north-east; north-north-west–south-south-east; all of 550–600 yards. A fuel tank of 4,000 gallons was installed and the tents used earlier gave way to some huts, one for the Flight CO's office, a watch office, a crew room and accommodation for a guard detachment of twelve to fifteen men. A Bellman hangar was erected in the north-east corner of the field, to supplement the Bessoneau that had been in the centre of the landing ground.

Gustav Hamel's Morane Saulnier at Wrexham in 1912/13.

Airspeed Feny G-ABSI with Alan Cobham's National Air Day, Borras, Wrexham on 25 September 1935.

A Hawker Audax in 1937. Many of these were to visit Borras from No.5 FTS, Sealand, as it was a relief landing ground for that station.

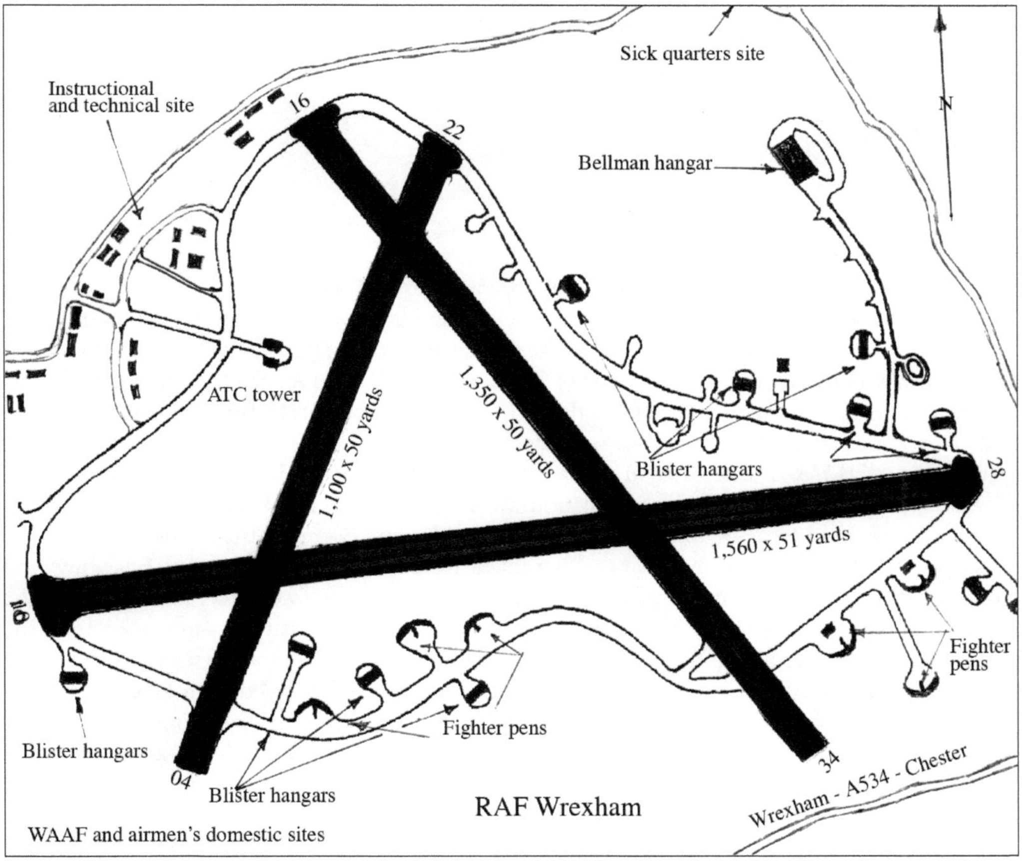

Map of RAF Wrexham showing runways and buildings. (I. Jones)

Opposite below: A Miles Master, also from No.5 FTS at Sealand. It is fair to say that pupil pilots from the base at Sealand spent more time flying around Wrexham and taking off or landing there than was done at the base. This was due to the boggy and crowded conditions at Sealand. (Via Aldon Ferguson)

No.9 Group Anti-Aircraft Co-operation Flight at Borras had several Mk IV Blenheims on strength in the spring of 1941.

A poor photograph but an important one. Defiant I-ZJ-H of No.96 Squadron at Wrexham in October 1941. The squadron's year-long conversion to the Beaufighter began here at the beginning of October 1941. (Via Don Clarke)

There was no control tower; the grass strip in wind was indicated with a Tee at the end of the runway, with another Tee on the signals square. All landing instructions were given by the aerodrome control pilot, armed with a Very pistol and an Aldis lamp. Positioning himself at the touchdown point of the runway in use, he would give a green for 'you may land' or red for 'wait and go around'.

No.7 OTU at Hawarden had been sending their pupils and aircraft to Borras. Their Hurricanes and Masters helped to fill the airspace over Borras until the shortage of Spitfire pilots dictated a change. No.7. OTU was disbanded on 1 November 1940 and No.57 OTU moved in with Spitfires. Hawarden was soggy and dangerous; the station was waiting for dry weather so that the contractors could move in and lay the essential hard runways. Again, it was Borras to the rescue.

There were frequent visits by other training units, mostly Tern Hill, Cranage and Shawbury, and many accidents and crashes occurred here, mostly involving the pupils of No.5 SFTS. One of the causes of crashes was the 'Borras Escarpement', which was an area of sharply rising ground making a 125ft break of slope, running north–south and forming the eastern and southern approaches to the airfield. This change of level was hard to spot from the air; many pilots failed to make the allowance and usually ended in a wheels-up landing. Some pilots likened it to landing on an aircraft carrier. One example of this occuring involved three Hurricanes of No.7 OTU, which all crashed at the same time in this area. They had been practicing dog-fighting and formation flying before heading for Borras at low level, possibly simulating ground attack on the RLG. L 1989 crashed wheels-up at Erlas Hall, L 1791 crashed at Brook Lawn and L 1870 came to ground at Lodge Farm. All three pilots were Belgian and survived the crashes. They had all failed to spot the change of ground level.

The Air Ministry had compulsory purchase orders out for the land required to make Borras a proper station. The Works Department was now in possession of Barras Lodge (104 acres), Borras Hall (56 acres), Borras Head (30 acres), Gourton Hall (2 acres), New Gorse (2 acres), The Gredington Arms, Llanypwll (8 acres), Wrexham Golf Club (55 acres) and sites on Plas Golbourne Farm. Plans were drawn and surveyors were on site but the building could not commence until December 1940. This left the various flying schools that were using Borras with seven more months for which to make other plans. The intention was to occupy the airfield with a night-fighter squadron.

No.5 SFTS moved from its base at Sealand to Tern Hill on 16 November 1940, just four weeks before the contractors moved in to Borras. The period from then until June 1941 saw McAlpine's men in occupation, 300 of them working by floodlight in the winter.

The commissioning of RAF Wrexham was to begin with the arrival, on 11 June 1941, of P/O D. Clark as equipment officer in charge of a small opening-up party. Permanent staff were posted in on 31 July. The first operational unit, No.9 Group Anti-Aircraft Co-operation Flight, comprising four pilots, thirty-seven ground crew and seven aircraft, flew in on 7 August 1941. They had formed at Speke on 5 May 1941 and were equipped with Blenheims IVs, Lysander IIs and Hurricane Is. Although No.5 SFTS was now at Tern Hill and had an improved airfield, it was still using Wrexham as if nothing had changed. From January to May 1941 the unit's Masters and Spitfires were again busy here. They had an establishment of no less than 150 Masters and a grass airfield that was flooded for weeks on end. In April and May even the Hurricanes of No.605 Squadron that were also based at Tern Hill used Wrexham.

So ended the scramble, partially at least, for Wrexham's landing field. On 21 January 1941, No.19 EFTS was formed at Sealand and used RAF Wrexham, as the previous tenant of Sealand – No.5 SFTS – had used the old Borras for their circuits and bumps. They had an establishment of seventy-two Tiger Moths. In April and May 1941 twenty-four of them were noted at Wrexham.

No.57 OTU from Hawarden continued to need the airfield. In fact, it had a Servicing and Maintenance party here on a permanent basis, under the command of Flt Lt D.H. Watkins. There was supposed to be a three-month period between opening up the station and the arrival of the resident No.96 Squadron, but the aircraft of that unit were using Wrexham repeatedly during that time. Since December 1940 No.96 Squadron had shared, as a lodger unit, the muddy grass

No.312 (Czech) Squadron had a presence at Wrexham during the spring and summer of 1941 as a forward base from Speke. Equipped with Hurricanes, they defended the north-west. (D.J. Smith)

No.96 Squadron arrived at Wrexham with Hurricanes as well as Defiants. In July 1941 the unit had sixteen Defiants and twelve Hurricanes. Some of the latter can be seen here at Wrexham, coded ZJ. (Via Don Clarke)

The Beaufighter IF of No.96 Squadron which was at Wrexham from October 1941 to October 1942. (Via Don Clarke)

Defiant Mk III target-tower at Wrexham with No.285 Squadron, photographed near the solitary Bellman hangar. (Via Don Clarke)

airfield of Cranage with No.2 School of Air Navigation. The attraction of the new hard runways at Wrexham, plus their night-flying facilities, proved too much and it is not surprising that the Defiants and Hurricanes used the place as a diversionary, a forward base or simply a convenience after a night sortie over Liverpool. The fact is that Wrexham was only half finished, yet eight of the squadron's aircraft were noted here in early May 1941, the month that nearly brought Liverpool to its knees. In July some sixteen Defiants and twelve Hurricanes of No.96 Squadron were logged as operating out of Wrexham, in addition to a couple of Hurricanes of No.151 Squadron that were based at Wittering. Speke's fighter squadrons used Wrexham as a forward base at various times. Nos 229, 315 (Polish) and 312 (Czech) Squadrons could be seen here prior to the official commissioning. The Blenheims of No.77 (Signals) Wing, which were based at Sefton Park, spent a period of time here, also prior to the opening and the official commissioning and it is believed that they were calibrating the radar site's equipment at Gwaunysgor, Gwespyr, Ruddlan and Llandudno in North Wales. Wrexham opened officially in June 1941, in No.9 Group of Fighter Command, but the advanced party of No.96 Squadron would not arrive before 20 October 1941, with twenty-one officers and 261 airmen. They were equipped with fourteen Defiants and two Hurricanes.

The operational life of No.96 Squadron at its previous base at Cranage had been productive, but there was little for them at Wrexham. The Luftwaffe had its work directed in the direction of Russia, so the squadron used this quiet spell to convert to Beaufighters, Mk I to start with and then Mk II, after doing a twin-engine course on Oxfords and a few Blenheims with dual

Miles Master II which equipped No.5 (P)AFU at Wrexham. (RAF Museum)

A No.577 Squadron Airspeed Oxford.

controls. The first Mk II arrived on 2 May 1942. When the unit had completed its conversion, it was moved to Honiley, in October 1942. Its only score while at this station was an errant barrage balloon. While this was going on another squadron was formed here, on 1 December. This was a redesignation of No.9 Group AACF, which now became No.285 Squadron, again in the anti-aircraft role, with seven Hudsons, four Blenheims, six Lysanders and a number of Defiants for target towing. The headquarters of this unit moved to Honiley with No.96 Squadron, leaving a detachment at Wrexham of one Oxford and four Defiants. The station became very quiet, with little in the way of aircraft movements, and at this time No.121 Airfield personnel moved into the living quarters for a few months.

On 1 March 1943 RAF Wrexham transferred from Fighter Command to No.21 Group (Spittalgate) Flying Training Command. The station now became the satellite to a succession of other stations, Cranage, Tern Hill and Calveley. First came No.5 (P) AFU (Advanced Flying Unit), which was previously a flying training school, but was redesignated on 11 April 1942 at Tern Hill. The school brought along its Masters Mk I and II. They stayed here until 4 May 1943. Next came No.17 (P) AFU, which had formed at Watton on 29 January 1942. This came to Wrexham from Calveley with over 170 Masters on establishment to share between them. The unit was to turn out several hundred pilots before disbanding in February 1944.

Another was No.11 (P) AFU that was received at Wrexham as a satellite to Calveley, a fleet of 130 Oxfords of 'B' Squadron, which had formerly been No.11 FTS at Shawbury. Then a change

Spitfire Mk16 of No.577 Squadron. This unit was at Wrexham for two years.

This building is sited at the eastern end of the main runway (28), near the fighter pens. It was No.96 Squadron's dispersal hut. This, apart from the hangar, is the only RAF building left. (I. Jones)

The Bellman hangar used by the machines of the Alfred McAlpine Co. (D.J. Smith)

RAF Wrexham now, submerged under the quarry. (Alfred McAlpine, via Don Clarke)

of policy saw No.11 (P) AFU convert to single-engine training and Masters were seen again at Wrexham. This unit disbanded on 1 June 1945 at Calveley.

No.5355 Airfield Construction Wing occupied the accommodation blocks to enable them to assemble plant and equipment in preparation for shipment overseas for the war against Japan. The only flying unit now was a detachment of No.577 Squadron, which formed at Castle Bromwich in December 1943 by merging elements of Nos 6, 7 and 8 Anti-Aircraft Co-operation Units. The detachment was equipped with Oxfords and Spitfire 16s and had been a lodger here since early 1943. It then departed to Atcham, where it stayed until the war's end, and eventually disbanded at Castle Bromwich on 15 June 1946.

RAF Wrexham now went on to a Care and Maintenance basis and the airfield was used to store armoured fighting vehicles under the name No.19 Fighting Vehicle Depot of the Royal Army Ordnance Corps. It was a shop window for the sale of tanks, half tracks and armoured cars until 22 October 1959 when it was put up for sale by auction, the lot comprising 255 acres with all the buildings and hangars.

The airfield went to the United Gravel Co., because of the gravel deposits under the airfield, for £23,000. This company was a subsidiary of Alfred McAlpine, which meant that the contractor that built RAF Wrexham now owned it. The company's corporate light aircraft used part of the main runway, but the rest of the airfield has been devoured, except for a few buildings that are used by the quarry workers, such as the Bellman hangar which shelters the company plant.

By the time the airfield closed the personnel numbered: ninety (RAF) and eight (WAAF) officers; 202 (RAF) and eight (WAAF) senior NCOs; 612 (RAF) and 240 (WAAF) other ranks.

The Airfield

There were three runways of concrete: west-north-west–east-south-east, 10/28, 1,560 yards; north-north-west–south-south-east, 16/34, 1,350 yards; north-east–south-west, 04/22, 1,100 yards.

There was one Bellman hangar and eleven Over Blister hangars, a control tower (drawing 12096/41) and six double fighter pens along the southern perimeter track. There were four accommodation sites, two communal sites and a sick quarters site.

Piper cubstrips at Wrexham – Acton Park Road and Acton Hall – not far from the Borras RAF accommodation sites. (Welsh Assembly Government Photographic Archive)

SIXTEEN

The US Army Airstrips in Wrexham 1943–45

There were no fewer than four US Army airstrips, all in the central and northern districts, also within a short distance of RAF Wrexham (Borras).

Grid reference SJ 344521, Acton Hall and Park

In February 1943 'B' Company of No.33 Signals Corps Construction Battalion arrived at Acton Hall and made preparations to receive the 400th Armoured Field Artillery Regiment in May, making the house their HQ. Soldiers were billeted in tents and huts in the park and were also housed in private homes all over Acton and the Garden Village. They were to do their training here until April 1944 when, typically, they departed quietly and suddenly.

Half a mile north of the hall was another cubstrip, now covered by housing on the Plas Golbourne Estate, at the junction of Box, Smithy and Barkers Lanes. This was probably used as a back up runway for the hall, in eventualities such as flooding or too much activity at one field or the other. Visiting the site now is pointless but it was necessary to make the attempt at least.

Grid reference SJ 346524, Beech House, Gresford

At Gresford, near the now abandoned colliery site, was Beeches House. Crossing the road (B5445) from the house is a large field, one mile west of what remains of the wartime RAF Wrexham, grid reference SJ 342538. It is believed this was used by the 322nd Field Artillery Battalion of the 83rd Infantry Division. Between the northern boundary of the RAF Wrexham site and the previously mentioned cubstrip, opposite the Beeches was yet another strip, just off the B5102 at the rear of the Pant-yr-Ochain House, grid reference SJ 358532. This was used by an unidentified unit, possibly the 733rd Field Artillery Battalion. It seems that this was replaced by a strip on RAF Wrexham Airfield, on the grass near the control tower, as shown on the top overhead photograph on page 171.

The Barkers Lane cubstrip has now been built on. This shot was taken in 2006, showing its approximate position, looking down the strip, with the hedge on the right and running along Barkers Lane. The crossroads are behind the photographer. (I. Jones)

The cubstrip at Acton Hall is now a public park, with new trees and shrubs diminishing the space (bottom of the overhead photograph). This view is looking south from what was the middle of the field. (I. Jones)

Looking north on the cubstrip mentioned on grid ref.SJ 358532, near Pant-yr-Ochain. Gresford is to the right. This field lies to the north-west of RAF Wrexham's technical site. This was the middle of the field. (I. Jones)

Looking south to the farms and houses bordering the airfield of RAF Wrexham. (I. Jones)

Facing north from the middle of the cubstrip opposite the Gresford Colliery site. (I. Jones)

Facing south on the cubstrip on the opposite side of the road to the Gresford Colliery site. Information on the locations of these cubstrips is the result of fine research by Mike Grant. (I. Jones)

Right: This overhead shows at the upper left the strip on grid reference SJ 358532. The US Army also used RAF Wrexham on a strip, as shown, near the control tower. This probably meant that the former strip was abandoned. (Welsh Assembly Government Photographic Archive)

Below: This overhead of 9 May 1946 shows Gresford Colliery at the top, with the cubstrip on the other side of the road (B5445). At the left of the picture is another sighting of the Barkers Lane strip. (Welsh Assembly Government Photographic Archive)

Main hangar with control tower (not active). The airport offices are in front. (I. Jones)

No.2 hangar at the Mid-Wales Airport, Welshpool. (I. Jones)

SEVENTEEN
Mid-Wales Airport

Grid reference SJ 210042, near the junction of the A483 and the A490 in Welshpool, Powys

Since 1986 a field at Welshpool has been in use as a grass airfield and the construction of a fine runway in 1990 has benefited the site so much that a second runway is imminent.

The main objective of this airport is for the businessman's convenience. Corporate flights depart from Welshpool to various airports throughout the UK and also to destinations in Europe. The company providing this service is called Awyr Cymru Cyf – Private Air Charter.

The airport came up for re-licensing in January 2003 and a second runway is now expected, as well as the expansion of many services such as a new, fixed-wing flying school using Cessna aircraft. At the moment these aircraft are in use as Fly-Hire, for self-use purposes. Also operating here are Wizard Helicopters, offering private and commercial pilot training and pleasure trips.

A lot of privately owned aircraft are based here and much use is made of the field by visitors to Wales from England. This will increase on the advent of another runway. The present runway, 22/04, is of concrete and asphalt, 2,723ft long and 59ft wide.

Welshpool town is well served by public transport. There is a regular train service from Shrewsbury and Birmingham in the east and Aberystwyth in the west. The airport is an hour away from the M6 and can easily be reached via the A483 and A490. This airport is in a lovely setting in the Vale of the Severn, with the river just to the east of the airfield.

At present the facilities of this fine airport, Mid-Wales Airport Ltd, are fuel – Avgas and Jet A1 – hangarage and catering. The hangar is brand new and large. When the expected expansion does take place it will allow larger aircraft to use the airport. They will then need appropriate air traffic control buildings and other safety devices to become an important regional airport.

The fine runway at Welshpool, Mid-Wales Airport. (I. Jones)

Touching down on 04 at Welshpool. (I. Jones)

Looking down to the south from 22 to 04 on the fine runway at Welshpool. (I. Jones)

A gala meet at the new Mid-Wales Airport on the outskirts of Welshpool. (I. Jones)

Port Dinorwic in August 1945 on the Menai Strait, from where the slate from Dinorwic Quarry was shipped to UK coastal towns and exported abroad. (Welsh Assembly Government Photographic Archive)

Dinorwic Quarry on the bank of Lake Padarn. The sheds shown were requisitioned by the Ministry of Aircraft Production for the use of NECACO. The tunnel is to the right at the end of the sheds. (Iris Morley-Jones, via Roy Sloan)

EIGHTEEN

The Slate Quarries of Gwynedd in the Air War

Grid reference SH 590610

For hundreds of years the major employer in this area of Gwynedd was the slate quarries and one of these was at Dinorwic, near Llanberis. Slates of every shape and size were mined and finished, then exported to all parts of the world as well as the UK from its own dedicated outlet at Port Dinorwic, some 8 miles to the north-west of the quarry, on the Menai Strait.

The quarry was a series of ledges and tunnels cut out of the mountain side on the northern side of Lake Padarn and the slate was brought from these quarry faces to a number of large factory sheds at the foot of the mountain. It was to be these sheds and a tunnel that were to be requisitioned by the Ministry of Aircraft Production, much to the annoyance of the owner, Vaynor Estates.

The Hunting Aircraft Co., with many contracts for aircraft components for its plant at Croydon, was suffering from air raids on two successive nights, bringing production to a halt. Although temporary arrangements were made to transfer some work to establishments in Epsom and Ewell it was vital to rebuild the factory at a location where it was safe from bombing.

Under the guidance of David Lloyd George, who was the leader of the opposition in Parliament at the time, the Minister of Aircraft Production (MAP), Lord Beaverbrook, toured the slate quarries of north-west Wales together with the local MP, Goronwy Owen, and Sir Percy Hunting, who was a friend of Lloyd George, in an attempt to find useful work for the unemployed of the area, and where the MAP was concerned, production space for Huntings. Included in the tour was a visit to Caernarfon, where a suitable facility for production was available at Peblig Mill. It was decided that this place and Dinorwic slate quarry would be very useful to Hunting Aircraft, and the sites were taken over on 28 August 1940 by a party of the staff, led by the manager Mr W.A. Summers, who was to be a fine leader of workers at all levels here.

Ironically, it had been planned earlier to build a plant at Newcastle and a company was formed entitled North-East Coast Aircraft Co. This was never activated; the Hunting Aircraft Co. was safer at Llanberis than Newcastle. The strange thing is that the name was retained – some say to confuse the enemy – and the facility at Dinorwic was to operate under the abbreviation NECACO. The company had to train local workers – people who had never seen the inside of a factory let alone operate machinery – in the use of tools, jigs and machines. Limited training was given at Caernarfon's Victoria Works on Victoria Docks, where some 3,000 trainees passed through the school.

The first of the many jigs used here arrived to make Defiant night-fighter wings. When things were ticking over well the company produced Stirling centre sections and bomb-bay floors, Halifax components, Wellington tail frames and bomb beams, and Lancaster front fuselages completely equipped. The machine shop was in a tunnel of the quarry workings, deep into the mountain side and 3,000ft below the summit. These were not ideal working conditions but after two years there were 3,000 people working for NECACO at Llanberis. The first deliveries of components were made in November 1940 and the last in 1945. The company then abandoned the site at the quarry but retained the Caernarfon factory for the post-war manufacturing of metal furniture. This ended in 1947 because of the difficulty in obtaining raw materials.

David Lloyd George (centre) with Dame Margeret Lloyd George and Percy Hunting during the negotiations for the establishment of the NECACO factory at Dinorwic. The gentleman seated next to Mr Lloyd George is W.A. Summers, the clever Works Manager of NECACO, Llanberis. (Via Roy Sloan)

Lancaster front fuselage production photographed in the old slate finishing shed at Dinorwic. (Via Reg Chambers-Jones)

An RAF overhead of Dinorwic on 10 May 1946. The lines on the mountain top look like water-gathering canals and perhaps the building at the centre is a pump-house. (Welsh Assembly Government Photographic Archive)

The factory was very successful and a real asset to the MAP as it was well run, especially when this was done with such unskilled workers. The site at Dinorwic stayed peaceful after the war until the Central Electricity Board built the Dinorwic Pumped Storage Power Station on the site.

No.31 Maintenance Unit, Llanberis Bomb Depot, Grid reference SH 570604

Another wartime user of the slate quarries was No.31 MU of the Royal Air Force. This was established in May 1941 on the southern side of Lake Padarn, on the opposite side of the lake to Dinorwic. The place was chosen for its deep chambers which had been left by the years of slate extraction, and with a small amount of civil engineering could be ideal for the storage of bombs for the operational squadrons of Bomber Command.

One of the assets in this choice was its proximity to an LMS Railway branch and main road, and it was officially opened on 2 June 1941 with Wing Commander R. Horstmann as CO. On the following day the first consignment arrived – ten rail trucks of bombs from RAF Swynnerton. By the end of the month 1,837 tons of bombs had been received and stored.

During the following months, when every available space had been used to store explosives, more dispersed sites were made available in the hamlets of Rhiwlas and Clwt-y-bont. The MU also expanded from the underground chambers onto the ledges of the mountainside.

Most of the 500 men onsite were civilians; the task of receiving, storing and delivering bombs, high explosives and incendiaries continued. The site had been purchased by the Air Ministry – 300 acres of land for £20,000. The buildings and railway sidings would bring it up to £470,000, but within this acreage were seven large, abandoned slate quarries which lay open to the sky. These had filled with water and were not suitable. One was dry, providing a space 300ft in diameter and the Air Ministry Works Directorate drew up plans to use it for a two-level, underfloor storage chamber. The lower floor received a 9in-thick ceiling, which was the ground for the chamber above. Railway access to the lower floor on two sidings, each with unloading platforms,

An overhead of No.31 MU, Llanberis, on 10 May 1946. The grid-like square near the A4086 – Caernarfon to Llanberis road – marked '1' on the photograph, is the now-roofless ground-floor bomb store that collapsed. '2' is the railroad and its curved entry into the bomb store. '3' is the flooded quarry-pit. (Welsh Assembly Government Photographic Archive)

Left: This oblique shot shows the upper part of the bomb depots area. Looking at the many roads and clearings from a north-easterly direction, it is hard to believe that this was taken in 1994 – fifty years on. (The Royal Commission on Ancient & Historic Monuments of Wales)

Opposite: No.31 MU, Llanberis, showing the eastern side of the depot. It was a large, widely dispersed bomb store after the troubles caused by the roof collapse. The area to the right of the shot held more roads, buildings and bomb dumps, and this was halfway up a mountain. (The Royal Commission on Ancient & Historical Monuments of Wales)

were built to provide two huge storage bays, one above the other. Three lifts were arranged to bring bombs from the lower to the upper levels, or the reverse when delivering to the airfields.

No.42 Group RAF, which was responsible for munitions and their storage, needed this depot very badly as Bomber Command was expanding and larger loads were being carried by each aircraft, but the Air Ministry stated that because of the shortage of materials for more important buildings than that at Llanberis, no priority could be given to No.31 MU. It was further stated that if the contractor was to carry on then he would have to go slowly and wait for every consignment of cement that could be released. What was even worse was that it was suggested that less cement be used in the mix and a reduction in the thickness of the floors and ceilings be carried out.

Building went on slowly through the summer and autumn of 1940, but had still not reached the first floor level by October. June 1941 saw the job completed and Mowlems, the contractor, was able to hand over the depot to No.42 Group.

The first consignment of bombs, consisting of ten railway wagons full, arrived at the depot, now named No.31 MU, from the Royal Ordnance Factory at Swynnerton on 3 June 1941. By the end of the month 332 truckloads had been received at the depot.

Each month the rate of delivery was increased or maintained; July and August saw 2,500 tons each, September 3,500 tons, October 6,000 tons, including the first consignment of 1,000lb bombs. No.42 Group needed more storage space, so another smaller facility was used at nearby Rhiwlas. Still searching for every square foot of space, the Air Ministry decided that it would be better to move all the non-explosive stores out of the underground galleries and spread them out onto the mountain side in dispersed batches, in makeshift shelters, making more room for the heavy bombs in the galleries.

At last No.31 MU was into its stride and did its job of receiving and then dispatching to operational squadrons, until the morning of 25 January when disaster struck! On that date a train of twenty-seven wagons, loaded with bombs from Swynnerton ROF was shunted into the low-level gallery's unloading bays. Deep grinding noises and grating of metal were heard by the men unloading. It was a noise quite different to the usual clanging of metal against metal as bombs rolled into one another. First dust then masonry began to fall in the bays and cracks appeared in the ceiling, opening up all the way along. The men working inside, mostly civilians, ran for their lives to the emergency exit and in seconds the ceiling broke away from the east wall at a point near the No.1 ammunition lift. It collapsed onto the train below, releasing thousands of bombs from the upper floor which rained down through the void. The walls folded inwards, like a house of cards, bombs from

above hitting bombs below. Fortunately none exploded and there were no casualties, but there were 18 tons of TNT in bulk and 75,000 tons of high-explosive bombs all trapped in the debris. The Air Ministry Works Directorate assessed the damage and came up with a recovery plan that was rushed through, as 15 per cent of Bomber Command's entire stock of HE bombs was in the wreckage.

An inquiry decided that the cause of the disaster was design faults compounded by poor concrete used in the construction. After removing the bombs and debris the area, now without a roof, would only be used for the storage of small arms ammunition from 25 June 1943, and roads would be built around the outside storage dispersals, each with its trench-like connotations like 'Petticoat Lane', Piccadilly Circus' and 'Rotten Row'. Buildings were erected on the hillside: station headquarters, a civilian canteen, standby set house, warden's office and detonator store.

At the war's end disposal of the munitions began. For instance, the entire stock of incendiary bombs at No.31 MU, some 4,500 tons, was deemed to be unusable due to prolonged storage in the open, and was destroyed here at the rate of sixty to seventy a day. In 1944–45 destruction of unserviceable ammunition on an extensive scale began. More and more bombs were discovered in pits and deep water that were displaced by the cave-in. Some could only arrive where they were found by human assistance and it appeared that No.31 MU had a job on its hands.

In 1955 No.31 MU was disbanded and the site came under the jurisdiction of RAF Llandwrog, a few miles to the west. This station also had a problem, as it was disposing of 70,000 TABUN nerve gas bombs from Germany, under the title Operation Sandcastle. The two places both had hazards for the communities around.

The troubles continued at Llanberis. The Royal Navy had divers searching the depths of the pits and lakes for explosive materials. The RAF maintained a section of bomb disposal experts at the site and the Royal Engineers had tasks at the old depot, cleaning up a dangerous place.

The whole site was fenced off from the public and the Air Ministry assumed that the worst was over, until a sixteen-year-old lad broke through the fence, found some kind of explosive ordnance which, while he was trying to dismantle it, blew up in his hands. This was in 1960 and it started yet another clean-up, leading to a process that went on until October 1975. Heartily fed-up, the Air Ministry attempted to foist the site onto Caernarfon County Council, suggesting that it could promote employment in the area. It was, of course, turned down and now belongs to the Crown, in perpetuity it seems.

Another of these slate quarries, at Manod in Blaenau Festiniog, was used to store the nation's art treasures during the war.

This overhead photograph tells the whole story: it is 8 July 1946 and No.78 MU has been disbanded for almost a year, yet the scrap aluminium and duralamin still surrounds the ex-tinplate works, except for the knoll to the east. (Central Registry of Air Photography for Wales, Welsh Assembly Government Photographic Archive)

NINETEEN
No.78 Maintenance Unit at Bynea

Grid reference SS 545992, near Llanelly Carms

Recent information relating to the airfields and landing grounds of West Wales
No.78 Maintenance Unit (MU) at Bynea is one of the sixteen aircraft salvage centres that were formed around the UK in No.43 Group, RAF, in 1939 and 1940. The MU was to be operated in its early history by a civilian contract handed to Messrs Macks Hauliers of Bradfield Road, Silvertown, London. This was to be based at the old Pencoed Tin-Plate Works and, comprising five sections, this establishment opened on 8 December 1940. On 22 March 1942 this MU became the service-manned Repair and Salvage Unit. From its initial beginnings, until becoming a wholly service unit, it had been parented by No.34 MU at Shrewsbury/Monkmoor.

There were no accommodation blocks at Bynea for the 200 or so servicemen and eighty WAAFs on station, they were all billeted out in homes in the surrounding area. Those who were married with homes in the area were allowed to go home each evening and there was an issue of a bicycle being available to each man or woman who needed one.

This MU was responsible for the recovery of all aircraft that had crashed or forced landed in western South Wales. No.32 MU at St Athan, which was another 'Crash and Smash', as they were known, operated eastern South Wales, although for both units the word 'south' was to be stretched northwards well into mid-Wales.

Most of the work for No.78 MU was the recovery of aircraft after sad mistakes in navigation, very often in poor weather conditions or at night, from training units all over the UK. They usually came down in remote, mountainous places and sometimes into the sea off the coast.

There were many US aircraft handled by No.78 MU. The south Atlantic ferry route from the US via Brazil, Ascension Islands and Marrakech to Valley brought many aircraft down around Wales. Liberators, Fortresses and Marauders, lost, short of fuel and in bad visibility, were to seek help. Some were saved but others crashed. Also German aircraft, either lost or damaged by British defences, would come to a sticky end in Wales.

One of the biggest jobs for No.78 MU was in June 1943. Sunderland flying boat T9114 from Pembroke Dock had suffered extensive hull damage while taking off in heavy seas after a brave rescue of downed aircrew in a dinghy in the Bay of Biscay. Due to the open hull it could not land back at Pembroke Dock, so the crew decided to attempt to land at the airfield of RAF Angle. This they did and the aircraft was repairable, so No.78 MU was called in. After fitting the beaching wheels that were used at Pembroke Dock for bringing the Sunderlands up the ramps to the hangars, the aircraft was towed to West Angle Bay, many fields and hedges over the hill from RAF Angle. It was repaired but was only used as an instructional airframe until scrapping.

On 10 March 1944 No.78 MU was called to a Marauder medium bomber of the USAAF which had crashed into the sea 250 yards offshore at Kenfig/Margam. This was another flight that had started at Marrakech and by the time No.78 MU arrived the crew had waded ashore. It was found possible to fix a line to the aircraft and tow it above the low water mark, using a rising tide.

This is the larger of the main buildings at what was – up to the mid-1930s – a tin-plate rolling mill. It had a useful gantry crane up in the roof and the front entrance to the MU can be seen. The current owner tells me that the asbestos corrugated-sheet shed by the side, in the middle of the picture, is the place where any bombs found at crash sites were to go. (I. Jones)

The Queen Mary articulated lorry was used extensively by No.78 MU.

The crash of Halifax 'F' of No.58 Squadron, Coastal Command, at St Davids. The aircraft just missed the hangar and tower, and was picked up by No.78 MU. (Via John Evans)

Complete salvage was carried out by No.78 MU: the engine bearers were cut and the engines were sent immediately to No.20 MU at Aston Down for dismantling and anti-corrosive action.

On 21 April 1944, at 9.15 p.m., another Marauder, 42-96277 of the USAAF, was reported as having crash-landed below high water mark on Pendine Sands. The aircraft was lifted and towed to a safe distance for dismantling, and was eventually disposed of as Category B. Both of these Marauders were on the last stages of ferry flights from the USA. Categories of damage were: Category U – undamaged; Category A – repairable on site by unit; Category Ac – repair beyond unit capacity; Category B – beyond repair on site; Category E – write-off.

By late 1943 and through 1944 many elements of the United States Army were based at Llanelly and its district, and among them was an engineer company that joined No.78 MU at the old Pencoed Tinplate Works. They were not in the aircraft recovery business, but that of road vehicle recovery and repair of the US Army jeeps, trucks and half tracks.

The RAF Training Stations in south-west Wales provided much of the work at Bynea. The regular crashes at Stormy Down Gunnery School and at Pembrey close by meant they were regular customers. The target-towing units and anti-aircraft school's Army co-operation aircraft were poorly maintained. With the best airframe and engine fitters at the operational squadrons there was a continual need for No.78 MU to find a route to some remote mountainside spots with no roads to make recoveries.

Elizabeth Williams was a WAAF driver at No.78 MU who did a four-year stint at Bynea and then a spell at St Athan, where she was to meet the man she eventually married. As a driver her task at Bynea was to take the officers to the crash sites. She saw so much horror – causes of crashes and the awful state of the casualties – that she vowed she would never fly. Long after her wartime service, whenever asked to go somewhere by air, she refused. Elizabeth died ten years ago and her wartime story was revealed to me by her niece, Diana Lomax.

The MU had a call-out on 9 May 1944. Two Ansons collided a mile offshore near Porthcawl and air-sea rescue boats from that port searched the sea for survivors in the hope that recovery of

No.7 Squadron Hampden, coded MG-M, serial L4042, of No.16 OTU, crashed in a landing overrun at Stormy Down in 1940, and was taken away by No.78 MU. (Via Philip Davies)

Fairwood Common on 9 April 1944. Halifax B Mk 2 series 1 of No.1658 Heavy Conversion Unit, based at Pocklington, crashed here onto WAAF site huts where many women were sleeping. One woman was killed and several injured. No.78 MU cleaned up the scrap. (*South Wales Evening Post*)

both aircraft could be achieved; however, both were failures. A dinghy and an item of wreckage were found. The next day an American Army diving unit and a salvage party from No.78 MU began the search for bodies. This lasted for nine days without success.

On 24 August 1944 a Liberator PB4Y of the US Navy took off from Dunkeswell, Devon, at 8.45 p.m. on a night familiarisation exercise that entailed a round trip of several airfields in the South West of England and West Wales. With a crew of six they made radio contact with St Eval, Cornwall, and then Fairwood Common obtained a fix on them with VHF. They seemed lost, as if the aircraft's use of instruments had failed. They should have been heading for RAF St Davids (Pembroke) but instead they were heading towards Llandovery. They passed very low over Glyntawe as they neared the mountains toward Sennybridge and then, at an altitude of 1,920ft, they hit the mountain Moel Feity, tragically just 10ft from its summit. The aircraft disintegrated, scattering burning fuel down the mountainside, where people as far away as Swansea could see the glow. All the crew died in this awful accident.

No.78 MU was called to the site, 6 miles to the south-west of Sennybridge. There was little to salvage but what there was in the way of secret instruments and weapons was gathered and the next day a clean-up was set in motion.

On 4 June 1943 a Maurauder of the USAAF, again on a ferry flight from Marrakech, crashed in thick fog on the slopes of Carn Llidi, which overlooks the Pembrokeshire beach of Whitesands Bay near St Davids. The pilot, Lt R.E. Lawrence, Lt H.H. Robertson, navigator, F/O J.G. Jackson, co-pilot, and Sgt W.A. Brown, engineer, were all were killed and the aircraft was written-off. This was yet another day's work for No.78 MU.

Another brutal crash was that of 9 April 1944. A Halifax B Mk 2 series 1 of No.1658 Heavy Conversion Unit, serial No.W7927, coded TT-R and based at Pocklington, crashed on RAF Fairwood Common's WAAF site at Upper Killay, killing Leading Aircraftwoman Dorothy Evans and injuring sixteen others while they slept. This aircraft was another write-off.

The Pencoed Tinplate Works was inactive before the Air Ministry commandeered the place. It provided a substantial indoor space in the largest shed, which was the rolling mill, and this provided the asset of an overhead electric gantry crane.

A few RAF huts were erected of Handcraft and Seco types for offices, a canteen and a guardroom at the gate, but little can be found now as the entire site is full of steel and iron scrap. One cannot even get to the main door that used to allow the articulated lorries in, bearing their loads so that the gantry crane could unload them.

The yard is now the property of Rees Metals. Mr R.R. Rees, the owner, was a youngster at Bynea when No.78 MU was in its heyday and now he and his son operate this site. I have been shown Halifax Main U/C wheels, jet engine parts and much more, but it was Mr Rees's memories that I really value.

One of my late contacts was Mr L.M. Roberts, who was with No.78 MU as an electrician. His job was to disconnect the wiring in the wings of a damaged aircraft, coil them up and identify them, then stuff them all into the fuselage to enable the wings to be removed from the fuselage. They were then loaded onto the Queen Mary. Unfortunately Mr Roberts was posted to India with only six months to go before demobilisation.

No.78 MU was disbanded on 31 August 1945. The personnel airmen and WAAF were posted off and the unit became just another wartime memory.

Sealand Farm on 8 September 1940. No work has been done to clear the ground yet. (Welsh Assembly Government Photographic Archive)

TWENTY

The Emergency Landing Ground at Sealand and Glan-y-Mor Farms

Grid reference SS 905720, three quarters of a mile west of Wick, 1 mile south of the satellite landing ground at St Brides and approximately 3 miles west of RAF Llandow's control tower.

Recent information relating to the landing grounds and airfields of South Wales

This large area of land, with the waters of the Bristol Channel forming its western limit, was surveyed in mid-1940 by the Air Ministry Works Department (AMWD) possibly for an even more important role than that of an ELG. St Brides had been surveyed, too, at that time and had been adopted as an SLG.

The oblique dated 8 September 1940 shows no changes to the fields or the Cwm (Valley), but as one explores the photograph dated 31 January 1942 there is evidence of hedge removal and in-filling of the hollows and ditches.

Andrew Thomas, the current farmer of Sealand, was not even born during the Second World War but knows the history of the place well. Similarly to Haroldston West ELG pastures, centuries old hedges, 6ft wide at the bases, made of stones and soil, with thick plants of old briar on top giving a height of up to 10ft, had been ripped out and levelled. All of these were replaced by lightweight fence posts, spaced further apart than was normal. The theory was that if an aircraft were to make an emergency landing that extended over several boundaries, the undercarriages would merely tear the posts out of the ground, without causing major damage to the aircraft. The farmers here, like the farmers at Haroldston West, complained that the ancient hedges now gone were not just boundaries but winter shelter from the wind for sheep and cattle.

The photograph at the top of page 190 shows the in-fill at the middle of the picture and also a serious attempt to fill the inland part of the Cwm. The photograph also shows anti-invasion posts running diagonally across some of the fields, at least that is what I believe them to be. If it appears ridiculous to clear a field for landing on one day and put posts in on the next, then one can only assume that there was a desperate need to get an aircraft down in nasty circumstances: the risk of damage to the aircraft by hitting the anti-invasion posts was a risk worth taking.

There were two options encouraged for landings from the sea direction on either side of the Cwm. The distance available from the cliff top to the St Brides to Marcross road that runs parallel with the coast is 1km. Landings from north to south or vice-versa were also possible as there were no high obstacles.

There was only one occasion when this emergency landing ground was used. On 8 December 1943, a USAAF B 24 H (Liberator) serial No.42-5313 of the 448th Bomb Group force-landed here with undercarriage up, finishing in the middle of the landing ground. The aircraft had left Marrakech after a transatlantic crossing at 2 a.m. It was destined for RAF St Mawgan, Cornwall, but in poor weather and with inaccurate navigation became lost. This was not the first time that aircraft on the Marrakech run had used the airfields of West Wales for emergency landings, but it was rare for East Wales.

Above: There is evidence of hedge removal on 31 January 1942 and in-filling is clear. The conventional lines of the old hedgerows have been replaced with wooden posts and wire. The diagonal lines are anti-invasion posts, to prevent landings by airborne troops. (Welsh Assembly Government Photographic Archive)

Above: This B24 is similar to the one that force-landed on Sealand Emergency Landing Ground.

Above: This is the view looking inland midway from the cliff, but this time on the southern side of the Cwm. (I. Jones)

Right: This is where the seaward approach brings you over the cliffs onto the Emergency Landing Ground at Sealand and Glan-y-Mor. (I. Jones)

Opposite below: This is the view when halfway along the 1km distance from the cliff top on the northern side of Cwm. (I. Jones)

No.3 Overseas Aircraft Preparation Unit was the major unit at Llandow at the time and it was their responsibility to get over to the crash site. It may be that it was Llandow's control tower that talked them down to the ELG. On arrival a policeman from the village of Wick attended the crash and found that none of the crew was hurt. They were transported to accommodation at RAF Llandow. It is rumoured that the policeman found the aircraft full of goodies: booze, cigarettes and chocolate bars!

The B24 was deemed Category E and was a write-off. After the usual inquiry into the reasons for the crash landing was completed, the crew were allowed to leave Llandow and on 17 December 2nd Lts L. Cepau, P. Lehman, O. Clavardoni and O. New, plus nine NCOs, left Llandow by rail to return to their duties.

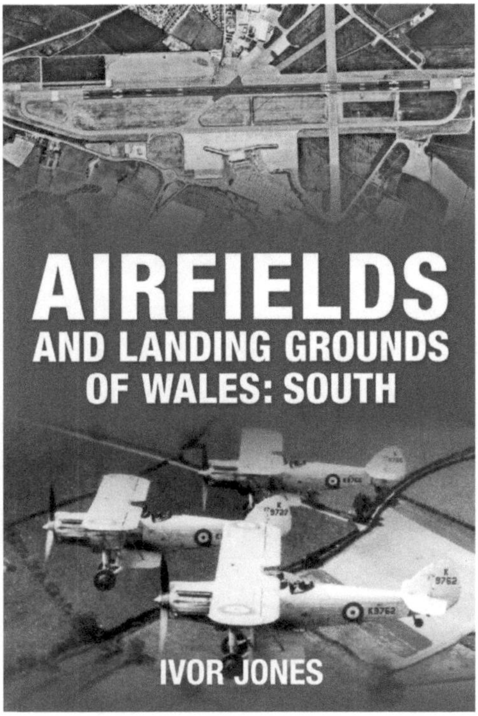

Airfields and Landing Grounds of Wales: South

IVOR JONES

Chronicling the location, history and fate of the many landing grounds, airfields and airports of South Wales from the early days of flight in 1911 through two world wars and up to the present day. Starting at the English border at Chepstow and finishing in Llandovery, Ivor Jones details such diverse histories as the meadows, parks and golf courses that were used as landing grounds by the US Army's liaison aircraft during the Secod World War, the RAF stations in the Vale of Glamorgan and the long-forgotten civil aerodromes of the 1930s that served to give flight to the average man.

ISBN: 978 0 7524 4273 0

£16.99

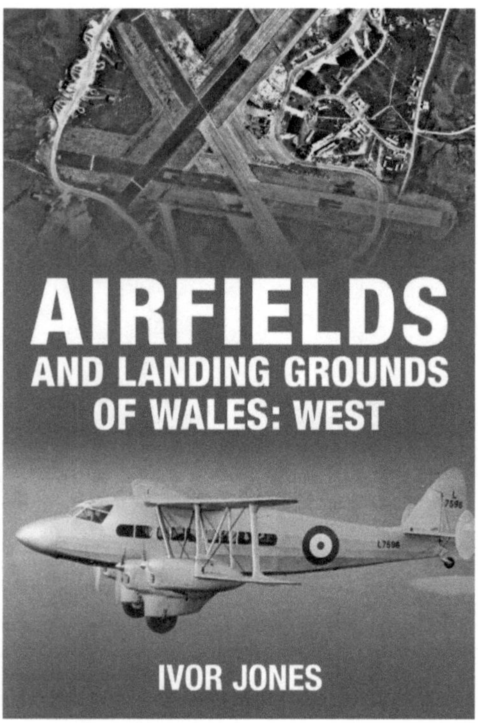

Airfields and Landing Grounds of Wales: West

IVOR JONES

In the second volume of the three-part series, Ivor Jones documents the West Wales region using a variety of archive photographs, overheads and maps. This area could be described as Coastal Command country, for that was the main contribution of the area to the history of Welsh wartime activity.

ISBN: 978 0 7524 4418 5

£16.99

If you are interested in purchasing other books published by The History Press, or in case you have difficulty finding any History Press books in your local bookshop, you can also place orders directly through our website

www.thehistorypress.co.uk